COMMON ERRORS IN STATISTICS (AND HOW TO AVOID THEM)

D0757510

COMMON ERRORS IN STATISTICS (AND HOW TO AVOID THEM)

Phillip I. Good
James W. Hardin

WILEY-INTERSCIENCE

A JOHN WILEY & SONS, INC., PUBLICATION

Copyright © 2003 by John Wiley & Sons, Inc. All rights reserved.

Published by John Wiley & Sons, Inc., Hoboken, New Jersey.
Published simultaneously in Canada.

No part of this publication may be reproduced, stored in a retrieval system, or transmitted in any form or by any means, electronic, mechanical, photocopying, recording, scanning, or otherwise, except as permitted under Section 107 or 108 of the 1976 United States Copyright Act, without either the prior written permission of the Publisher, or authorization through payment of the appropriate per-copy fee to the Copyright Clearance Center, Inc., 222 Rosewood Drive, Danvers, MA 01923, 978-750-8400, fax 978-750-4470, or on the web at www.copyright.com. Requests to the Publisher for permission should be addressed to the Permissions Department, John Wiley & Sons, Inc., 111 River Street, Hoboken, NJ 07030, (201) 748-6011, fax (201) 748-6008, e-mail: permreq@wiley.com.

Limit of Liability/Disclaimer of Warranty: While the publisher and author have used their best efforts in preparing this book, they make no representations or warranties with respect to the accuracy or completeness of the contents of this book and specifically disclaim any implied warranties of merchantability or fitness for a particular purpose. No warranty may be created or extended by sales representatives or written sales materials. The advice and strategies contained herein may not be suitable for your situation. You should consult with a professional where appropriate. Neither the publisher nor author shall be liable for any loss of profit or any other commercial damages, including but not limited to special, incidental, consequential, or other damages.

For general information on our other products and services please contact our Customer Care Department within the U.S. at 877-762-2974, outside the U.S. at 317-572-3993 or fax 317-572-4002.

Wiley also publishes its books in a variety of electronic formats. Some content that appears in print, however, may not be available in electronic format.

Library of Congress Cataloging-in-Publication Data:

Good, Phillip I.
    Common errors in statistics (and how to avoid them)/Phillip I. Good, James W. Hardin.
        p. cm.
    Includes bibliographical references and index.
    ISBN 0-471-46068-0 (pbk. : acid-free paper)
    1. Statistics. I. Hardin, James W. (James William) II. Title.

QA276.G586 2003
519.5—dc21

                                                2003043279

Printed in the United States of America.

10 9 8 7 6 5 4 3

Contents

Preface

ONE OF THE VERY FIRST STATISTICAL APPLICATIONS ON which Dr. Good worked was an analysis of leukemia cases in Hiroshima, Japan following World War II; on August 7, 1945 this city was the target site of the first atomic bomb dropped by the United States. Was the high incidence of leukemia cases among survivors the result of exposure to radiation from the atomic bomb? Was there a relationship between the number of leukemia cases and the number of survivors at certain distances from the atomic bomb's epicenter?

To assist in the analysis, Dr. Good had an electric (not an electronic) calculator, reams of paper on which to write down intermediate results, and a prepublication copy of Scheffe's *Analysis of Variance*. The work took several months and the results were somewhat inconclusive, mainly because he could never seem to get the same answer twice—a consequence of errors in transcription rather than the absence of any actual relationship between radiation and leukemia.

Today, of course, we have high-speed computers and prepackaged statistical routines to perform the necessary calculations. Yet, statistical software will no more make one a statistician than would a scalpel turn one into a neurosurgeon. Allowing these tools to do our thinking for us is a sure recipe for disaster.

Pressed by management or the need for funding, too many research workers have no choice but to go forward with data analysis regardless of the extent of their statistical training. Alas, while a semester or two of undergraduate statistics may suffice to develop familiarity with the names of some statistical methods, it is not enough to be aware of all the circumstances under which these methods may be applicable.

The purpose of the present text is to provide a mathematically rigorous but readily understandable foundation for statistical procedures. Here for the second time are such basic concepts in statistics as null and alternative hypotheses, p value, significance level, and power. Assisted by reprints from the statistical literature, we reexamine sample selection, linear regression, the analysis of variance, maximum likelihood, Bayes' Theorem, meta-analysis, and the bootstrap.

Now the good news: Dr. Good's articles on women's sports have appeared in the *San Francisco Examiner, Sports Now,* and *Volleyball Monthly.* So, if you can read the sports page, you'll find this text easy to read and to follow. Lest the statisticians among you believe this book is too introductory, we point out the existence of hundreds of citations in statistical literature calling for the comprehensive treatment we have provided. Regardless of past training or current specialization, this book will serve as a useful reference; you will find applications for the information contained herein whether you are a practicing statistician or a well-trained scientist who just happens to apply statistics in the pursuit of other science.

The primary objective of the opening chapter is to describe the main sources of error and provide a preliminary prescription for avoiding them. The hypothesis formulation—data gathering—hypothesis testing and estimate cycle is introduced, and the rationale for gathering additional data before attempting to test after-the-fact hypotheses is detailed.

Chapter 2 places our work in the context of decision theory. We emphasize the importance of providing an interpretation of each and every potential outcome in advance of consideration of actual data.

Chapter 3 focuses on study design and data collection for failure at the planning stage can render all further efforts valueless. The work of Vance Berger and his colleagues on selection bias is given particular emphasis.

Desirable features of point and interval estimates are detailed in Chapter 4 along with procedures for deriving estimates in a variety of practical situations. This chapter also serves to debunk several myths surrounding estimation procedures.

Chapter 5 reexamines the assumptions underlying testing hypotheses. We review the impacts of violations of assumptions, and we detail the procedures to follow when making two- and k-sample comparisons. In addition, we cover the procedures for analyzing contingency tables and two-way experimental designs if standard assumptions are violated.

Chapter 6 is devoted to the value and limitations of Bayes' Theorem, meta-analysis, and resampling methods.

Chapter 7 lists the essentials of any report that will utilize statistics, debunks the myth of the "standard" error, and describes the value and limitations of p values and confidence intervals for reporting results. Practical significance is distinguished from statistical significance, and induction is distinguished from deduction.

Twelve rules for more effective graphic presentations are given in Chapter 8 along with numerous examples of the right and wrong ways to maintain reader interest while communicating essential statistical information.

Chapters 9 through 11 are devoted to model building and to the assumptions and limitations of standard regression methods and data mining techniques. A distinction is drawn between goodness of fit and prediction, and the importance of model validation is emphasized. Seminal articles by David Freedman and Gail Gong are reprinted.

Finally, for the further convenience of readers, we provide a glossary grouped by related but contrasting terms, a bibliography, and subject and author indexes.

Our thanks to William Anderson, Leonardo Auslender, Vance Berger, Peter Bruce, Bernard Choi, Tony DuSoir, Cliff Lunneborg, Mona Hardin, Gunter Hartel, Fortunato Pesarin, Henrik Schmiediche, Marjorie Stinespring, and Peter A. Wright for their critical reviews of portions of this text. Doug Altman, Mark Hearnden, Elaine Hand, and David Parkhurst gave us a running start with their bibliographies.

We hope you soon put this textbook to practical use.

Phillip Good
Huntington Beach, CA
brother_unknown@yahoo.com

James Hardin
College Station, TX
jhardin@stat.tamu.edu

Part I
FOUNDATIONS

"Don't think—use the computer."
G. Dyke

Chapter 1

Sources of Error

STATISTICAL PROCEDURES FOR HYPOTHESIS TESTING, ESTIMATION, AND MODEL building are only a *part* of the decision-making process. They should never be quoted as the sole basis for making a decision (yes, even those procedures that are based on a solid deductive mathematical foundation). As philosophers have known for centuries, extrapolation from a sample or samples to a larger incompletely examined population must entail a leap of faith.

The sources of error in applying statistical procedures are legion and include all of the following:

- Using the same set of data both to formulate hypotheses and to test them.
- Taking samples from the wrong population or failing to specify the population(s) about which inferences are to be made in advance.
- Failing to draw random, representative samples.
- Measuring the wrong variables or failing to measure what you'd hoped to measure.
- Using inappropriate or inefficient statistical methods.
- Failing to validate models.

But perhaps the most serious source of error lies in letting statistical procedures make decisions for you.

In this chapter, as throughout this text, we offer first a preventive prescription, followed by a list of common errors. If these prescriptions are followed carefully, you will be guided to the correct, proper, and effective use of statistics and avoid the pitfalls.

Common Errors in Statistics (and How to Avoid Them), by Phillip I. Good and James W. Hardin.
ISBN 0-471-46068-0   Copyright © 2003 John Wiley & Sons, Inc.

PRESCRIPTION

Statistical methods used for experimental design and analysis should be viewed in their rightful role as merely a part, albeit an essential part, of the decision-making procedure.

Here is a partial prescription for the error-free application of statistics.

1. Set forth your objectives and the use you plan to make of your research *before* you conduct a laboratory experiment, a clinical trial, or survey and *before* you analyze an existing set of data.

2. Define the population to which you will apply the results of your analysis.

3. List all possible sources of variation. Control them or measure them to avoid their being confounded with relationships among those items that are of primary interest.

4. Formulate your hypothesis and all of the associated alternatives. (See Chapter 2.) List possible experimental findings along with the conclusions you would draw and the actions you would take if this or another result should prove to be the case. Do all of these things *before* you complete a single data collection form and *before* you turn on your computer.

5. Describe in detail how you intend to draw a representative sample from the population. (See Chapter 3.)

6. Use estimators that are impartial, consistent, efficient, and robust and that involve minimum loss. (See Chapter 4.) To improve results, focus on sufficient statistics, pivotal statistics, and admissible statistics, and use interval estimates. (See Chapters 4 and 5.)

7. Know the assumptions that underlie the tests you use. Use those tests that require the minimum of assumptions and are most powerful against the alternatives of interest. (See Chapter 5.)

8. Incorporate in your reports the complete details of how the sample was drawn and describe the population from which it was drawn. If data are missing or the sampling plan was not followed, explain why and list all differences between data that were present in the sample and data that were missing or excluded. (See Chapter 7.)

FUNDAMENTAL CONCEPTS

Three concepts are fundamental to the design of experiments and surveys: variation, population, and sample.

A thorough understanding of these concepts will forestall many errors in the collection and interpretation of data.

If there were no variation, if every observation were predictable, a mere repetition of what had gone before, there would be no need for statistics.

Variation

Variation is inherent in virtually all our observations. We would not expect outcomes of two consecutive spins of a roulette wheel to be identical. One result might be red, the other black. The outcome varies from spin to spin.

There are gamblers who watch and record the spins of a single roulette wheel hour after hour hoping to discern a pattern. A roulette wheel is, after all, a mechanical device and perhaps a pattern will emerge. But even those observers do not anticipate finding a pattern that is 100% deterministic. The outcomes are just too variable.

Anyone who spends time in a schoolroom, as a parent or as a child, can see the vast differences among individuals. This one is tall, today, that one short. Half an aspirin and Dr. Good's headache is gone, but his wife requires four times that dosage.

There is variability even among observations on deterministic formula-satisfying phenomena such as the position of a planet in space or the volume of gas at a given temperature and pressure. Position and volume satisfy Kepler's Laws and Boyle's Law, respectively, but the observations we collect will depend upon the measuring instrument (which may be affected by the surrounding environment) and the observer. Cut a length of string and measure it three times. Do you record the same length each time?

In designing an experiment or survey, we must always consider the possibility of errors arising from the measuring instrument and from the observer. It is one of the wonders of science that Kepler was able to formulate his laws at all, given the relatively crude instruments at his disposal.

Population

The population(s) of interest must be clearly defined before we begin to gather data.

From time to time, someone will ask us how to generate confidence intervals (see Chapter 7) for the statistics arising from a total census of a population. Our answer is no, we cannot help. Population statistics (mean, median, 30th percentile) are not estimates. They are fixed values and will be known with 100% accuracy if two criteria are fulfilled:

1. Every member of the population is observed.
2. All the observations are recorded correctly.

Confidence intervals would be appropriate if the first criterion is violated, because then we are looking at a sample, not a population. And if the second criterion is violated, then we might want to talk about the confidence we have in our measurements.

Debates about the accuracy of the 2000 United States Census arose from doubts about the fulfillment of these criteria.[1] "You didn't count the homeless," was one challenge. "You didn't verify the answers," was another. Whether we collect data for a sample or an entire population, both these challenges or their equivalents can and should be made.

Kepler's "laws" of planetary movement are not testable by statistical means when applied to the original planets (Jupiter, Mars, Mercury, and Venus) for which they were formulated. But when we make statements such as "Planets that revolve around Alpha Centauri will also follow Kepler's Laws," then we begin to view our original population, the planets of our sun, as a sample of all possible planets in all possible solar systems.

A major problem with many studies is that the population of interest is not adequately defined before the sample is drawn. Don't make this mistake. A second major source of error is that the sample proves to have been drawn from a different population than was originally envisioned. We consider this problem in the next section and again in Chapters 2, 5, and 6.

Sample

A sample is any (proper) subset of a population.

Small samples may give a distorted view of the population. For example, if a minority group comprises 10% or less of a population, a jury of 12 persons selected at random from that population fails to contain any members of that minority at least 28% of the time.

As a sample grows larger, or as we combine more clusters within a single sample, the sample will grow to more closely resemble the population from which it is drawn.

How large a sample must be to obtain a sufficient degree of closeness will depend upon the manner in which the sample is chosen from the population. Are the elements of the sample drawn at random, so that each unit in the population has an equal probability of being selected? Are the elements of the sample drawn independently of one another?

If either of these criteria is not satisfied, then even a very large sample may bear little or no relation to the population from which it was drawn.

An obvious example is the use of recruits from a Marine boot camp as representatives of the population as a whole or even as representatives of all Marines. In fact, any group or cluster of individuals who live, work, study, or pray together may fail to be representative for any or all of the following reasons (Cummings and Koepsell, 2002):

---

[1] City of New York v. Department of Commerce, 822 F. Supp. 906 (E.D.N.Y, 1993). The arguments of four statistical experts who testified in the case may be found in Volume 34 of *Jurimetrics*, 1993, 64–115.

1. Shared exposure to the same physical or social environment
2. Self-selection in belonging to the group
3. Sharing of behaviors, ideas, or diseases among members of the group

A sample consisting of the first few animals to be removed from a cage will not satisfy these criteria either, because, depending on how we grab, we are more likely to select more active or more passive animals. Activity tends to be associated with higher levels of corticosteroids, and corticosteroids are associated with virtually every body function.

Sample bias is a danger in every research field. For example, Bothun [1998] documents the many factors that can bias sample selection in astronomical research.

To forestall sample bias in your studies, determine before you begin the factors can affect the study outcome (gender and life style, for example). Subdivide the population into strata (males, females, city dwellers, farmers) and then draw separate samples from each stratum. Ideally, you would assign a random number to each member of the stratum and let a computer's random number generator determine which members are to be included in the sample.

Surveys and Long-Term Studies

Being selected at random does not mean that an individual will be willing to participate in a public opinion poll or some other survey. But if survey results are to be representative of the population at large, then pollsters must find some way to interview nonresponders as well. This difficulty is only exacerbated in long-term studies, because subjects fail to return for follow-up appointments and move without leaving a forwarding address. Again, if the sample results are to be representative, some way must be found to report on subsamples of the nonresponders and the dropouts.

AD HOC, POST HOC HYPOTHESES

Formulate and write down your hypotheses before you examine the data.

Patterns in data can suggest, but cannot confirm hypotheses unless these hypotheses were formulated *before* the data were collected.

Everywhere we look, there are patterns. In fact, the harder we look, the more patterns we see. Three rock stars die in a given year. Fold the United States 20-dollar bill in just the right way and not only the Pentagon but the Twin Towers in flames are revealed. It is natural for us to want to attribute some underlying cause to these patterns. But those who have studied the laws of probability tell us that more often than not patterns are simply the result of random events.

Put another way, finding at least one cluster of events in time or in space has a greater probability than finding no clusters at all (equally spaced events).

How can we determine whether an observed association represents an underlying cause and effect relationship or is merely the result of chance? The answer lies in our research protocol. When we set out to test a specific hypothesis, the probability of a specific event is predetermined. But when we uncover an apparent association, one that may well have arisen purely by chance, we cannot be sure of the association's validity until we conduct a second set of controlled trials.

In the International Study of Infarct Survival [1988], patients born under the Gemini or Libra astrological birth signs did not survive as long when their treatment included aspirin. By contrast, aspirin offered apparent beneficial effects (longer survival time) to study participants from all other astrological birth signs.

Except for those who guide their lives by the stars, there is no hidden meaning or conspiracy in this result. When we describe a test as significant at the 5% or 1-in-20 level, we mean that 1 in 20 times we'll get a significant result even though the hypothesis is true. That is, when we test to see if there are any differences in the baseline values of the control and treatment groups, if we've made 20 different measurements, we can expect to see at least one statistically significant difference; in fact, we will see this result almost two-thirds of the time. This difference will not represent a flaw in our design but simply chance at work. To avoid this undesirable result—that is, to avoid attributing statistical significance to an insignificant random event, a so-called Type I error—we must distinguish between the hypotheses with which we began the study and those that came to mind afterward. We must accept or reject these hypotheses at the original significance level while demanding additional corroborating evidence for those exceptional results (such as a dependence of an outcome on astrological sign) that are uncovered for the first time during the trials.

No reputable scientist would ever report results before successfully reproducing the experimental findings twice, once in the original laboratory and once in that of a colleague.[2] The latter experiment can be particularly telling, because all too often some overlooked factor not controlled in the experiment—such as the quality of the laboratory water—proves responsible for the results observed initially. It is better to be found wrong

---

[2] Remember "cold fusion?" In 1989, two University of Utah professors told the newspapers that they could fuse deuterium molecules in the laboratory, solving the world's energy problems for years to come. Alas, neither those professors nor anyone else could replicate their findings, though true believers abound, http://www.ncas.org/erab/intro.htm.

in private than in public. The only remedy is to attempt to replicate the findings with different sets of subjects, replicate, and then replicate again.

Persi Diaconis [1978] spent some years as a statistician investigating paranormal phenomena. His scientific inquiries included investigating the powers linked to Uri Geller, the man who claimed he could bend spoons with his mind. Diaconis was not surprised to find that the hidden "powers" of Geller were more or less those of the average nightclub magician, down to and including forcing a card and taking advantage of ad hoc, post hoc hypotheses.

When three buses show up at your stop simultaneously, or three rock stars die in the same year, or a stand of cherry trees is found amid a forest of oaks, a good statistician remembers the Poisson distribution. This distribution applies to relatively rare events that occur independently of one another. The calculations performed by Siméon-Denis Poisson reveal that if there is an average of one event per interval (in time or in space), then while more than one-third of the intervals will be empty, at least one-fourth of the intervals are likely to include multiple events.

Anyone who has played poker will concede that one out of every two hands contains "something" interesting. Don't allow naturally occurring results to fool you or to lead you to fool others by shouting, "Isn't this incredible?"

The purpose of a recent set of clinical trials was to see if blood flow and distribution in the lower leg could be improved by carrying out a simple surgical procedure prior to the administration of standard prescription medicine.

The results were disappointing on the whole, but one of the marketing representatives noted that the long-term prognosis was excellent when a marked increase in blood flow was observed just after surgery. She suggested we calculate a p value[3] for a comparison of patients with an improved blood flow versus patients who had taken the prescription medicine alone.

Such a p value would be meaningless. Only one of the two samples of patients in question had been taken at random from the population (those patients who received the prescription medicine alone). The other sample (those patients who had increased blood flow following surgery) was determined after the fact. In order to extrapolate results from the samples in hand to a larger population, the samples must be taken at random from, and be representative of, that population.

---

[3] A p value is the probability under the primary hypothesis of observing the set of observations we have in hand. We can calculate a p value once we make a series of assumptions about how the data were gathered. These days, statistical software does the calculations, but its still up to us to verify that the assumptions are correct.

The preliminary findings clearly called for an examination of surgical procedures and of patient characteristics that might help forecast successful surgery. But the generation of a p value and the drawing of any final conclusions had to wait on clinical trials specifically designed for that purpose.

This doesn't mean that one should not report anomalies and other unexpected findings. Rather, one should not attempt to provide p values or confidence intervals in support of them. Successful researchers engage in a cycle of theorizing and experimentation so that the results of one experiment become the basis for the hypotheses tested in the next.

A related, extremely common error whose resolution we discuss at length in Chapters 10 and 11 is to use the same data to select variables for inclusion in a model and to assess their significance. Successful model builders develop their frameworks in a series of stages, validating each model against a second independent data set before drawing conclusions.

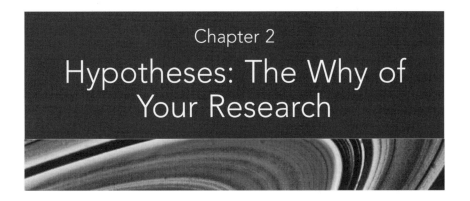

Chapter 2

Hypotheses: The Why of Your Research

IN THIS CHAPTER WE REVIEW HOW TO FORMULATE a hypothesis that is testable by statistical means, the appropriate use of the null hypothesis, Neyman–Pearson theory, the two types of error, and the more general theory of decisions and losses.

PRESCRIPTION

Statistical methods used for experimental design and analysis should be viewed in their rightful role as merely a part, albeit an essential part, of the decision-making procedure.

1. Set forth your objectives and the use you plan to make of your research *before* you conduct a laboratory experiment, a clinical trial, or a survey and *before* you analyze an existing set of data.

2. Formulate your hypothesis and all of the associated alternatives. List possible experimental findings along with the conclusions you would draw and the actions you would take if this or another result should prove to be the case. Do all of these things *before* you complete a single data collection form and *before* you turn on your computer.

WHAT IS A HYPOTHESIS?

A well-formulated hypothesis will be both quantifiable and testable—that is, involve measurable quantities or refer to items that may be assigned to mutually exclusive categories.

A well-formulated statistical hypothesis takes one of the following forms: "Some measurable characteristic of a population takes one of a specific set

Common Errors in Statistics (and How to Avoid Them), by Phillip I. Good and James W. Hardin. ISBN 0-471-46068-0   Copyright © 2003 John Wiley & Sons, Inc.

of values." or "Some measurable characteristic takes different values in different populations, the difference(s) taking a specific pattern or a specific set of values."

Examples of well-formed statistical hypotheses include the following:

- "For males over 40 suffering from chronic hypertension, a 100 mg daily dose of this new drug lowers diastolic blood pressure an average of 10 mm Hg."
- "For males over 40 suffering from chronic hypertension, a daily dose of 100 mg of this new drug lowers diastolic blood pressure an average of 10 mm Hg more than an equivalent dose of metoprolol."
- "Given less than 2 hours per day of sunlight, applying from 1 to 10 lb of 23–2–4 fertilizer per 1000 square feet will have no effect on the growth of fescues and Bermuda grasses."

"All redheads are passionate" is not a well-formed statistical hypothesis—not merely because "passionate" is ill-defined, but because the word "All" indicates that the phenomenon is not statistical in nature.

Similarly, logical assertions of the form "Not all," "None," or "Some" are not statistical in nature. The restatement, "80% of redheads are passionate," would remove this latter objection.

The restatements, "Doris J. is passionate," or "Both Good brothers are 5′10″ tall," also are not statistical in nature because they concern specific individuals rather than populations (Hagood, 1941).

If we quantify "passionate" to mean "has an orgasm more than 95% of the time consensual sex is performed," then the hypothesis "80% of redheads are passionate" becomes testable. Note that defining "passionate" to mean "has an orgasm every time consensual sex is performed" would not be provable as it is a statement of the "all or none" variety.

Finally, note that until someone succeeds in locating unicorns, the hypothesis "80% of unicorns are passionate" is *not* testable.

Formulate your hypotheses so they are quantifiable, testable, and statistical in nature.

How Precise Must a Hypothesis Be?

The chief executive of a drug company may well express a desire to test whether "our anti-hypertensive drug can beat the competition." But to apply statistical methods, a researcher will need precision on the order of "For males over 40 suffering from chronic hypertension, a daily dose of 100 mg of our new drug will lower diastolic blood pressure an average of 10 mm Hg more than an equivalent dose of metoprolol."

The researcher may want to test a preliminary hypothesis on the order of "For males over 40 suffering from chronic hypertension, there is a daily dose of our new drug which will lower diastolic blood pressure an average of 20 mm Hg." But this hypothesis is imprecise. What if the necessary dose of the new drug required taking a tablet every hour? Or caused liver malfunction? Or even death? First, the researcher would conduct a set of clinical trials to determine the maximum tolerable dose (MTD) and then test the hypothesis, "For males over 40 suffering from chronic hypertension, a daily dose of one-third to one-fourth the MTD of our new drug will lower diastolic blood pressure an average of 20 mm Hg."

A BILL OF RIGHTS

- Scientists can and should be encouraged to make subgroup analyses.
- Physicians and engineers should be encouraged to make decisions utilizing the findings of such analyses.
- Statisticians and other data analysts can and should rightly refuse to give their imprimatur to related tests of significance.

In a series of articles by Horwitz et al. [1998], a physician and his colleagues strongly criticize the statistical community for denying them (or so they perceive) the right to provide a statistical analysis for subgroups not contemplated in the original study protocol. For example, suppose that in a study of the health of Marine recruits, we notice that not one of the dozen or so women who received the vaccine contracted pneumonia. Are we free to provide a p value for this result?

Statisticians Smith and Egger [1998] argue against hypothesis tests of subgroups chosen after the fact, suggesting that the results are often likely to be explained by the "play of chance." Altman [1998b, pp. 301–303], another statistician, concurs noting that ". . . the observed treatment effect is expected to vary across subgroups of the data . . . simply through chance variation" and that "doctors seem able to find a biologically plausible explanation for any finding." This leads Horwitz et al. [1998] to the incorrect conclusion that Altman proposes we "dispense with clinical biology (biologic evidence and pathophysiologic reasoning) as a basis for forming subgroups." Neither Altman nor any other statistician would quarrel with Horwitz et al.'s assertion that physicians must investigate "how do we [physicians] do our best for a particular patient."

Scientists can and should be encouraged to make subgroup analyses. Physicians and engineers should be encouraged to make decisions based upon them. Few would deny that in an emergency, satisficing [coming up with workable, fast-acting solutions without complete information] is better than optimizing.[1] But, by the same token, statisticians should not

---

[1] Chiles [2001, p. 61].

be pressured to give their imprimatur to what, in statistical terms, is clearly an improper procedure, nor should statisticians mislabel suboptimal procedures as the best that can be done.[2]

We concur with Anscombe [1963], who writes, ". . . the concept of error probabilities of the first and second kinds . . . has no direct relevance to experimentation. . . . The formation of opinions, decisions concerning further experimentation and other required actions, are not dictated . . . by the formal analysis of the experiment, but call for judgment and imagination. . . . It is unwise for the experimenter to view himself seriously as a decision-maker. . . . The experimenter pays the piper and calls the tune he likes best; but the music is broadcast so that others might listen. . . ."

NULL HYPOTHESIS

"A major research failing seems to be the exploration of uninteresting or even trivial questions. . . . In the 347 sampled articles in *Ecology* containing null hypotheses tests, we found few examples of null hypotheses that seemed biologically plausible." Anderson, Burnham, and Thompson [2000].

Test Only Relevant Null Hypotheses

The "null hypothesis" has taken on an almost mythic role in contemporary statistics. Obsession with the "null" has been allowed to shape the direction of our research. We've let the tool use us instead of our using the tool.[3]

While a null hypothesis can facilitate statistical inquiry—an exact permutation test is impossible without it—it is never mandated. In any event, virtually any quantifiable hypothesis can be converted into null form. There is no excuse and no need to be content with a meaningless null.

To test that the mean value of a given characteristic is three, subtract three from each observation and then test the "null hypothesis" that the mean value is zero.

Often, we want to test that the size of some effect is inconsequential, not zero but close to it, smaller than d, say, where d is the smallest biological, medical, physical or socially relevant effect in your area of research. Again, subtract d from each observation, before proceeding to test a null hypothesis. In Chapter 5 we discuss an alternative approach using confidence intervals for tests of equivalence.

---

[2] One is reminded of the Dean, several of them in fact, who asked me to alter my grades. "But that is something you can do as easily as I." "Why Dr. Good, I would never dream of overruling one of my instructors."

[3] See, for example, Hertwig and Todd [2000].

To test that "80% of redheads are passionate," we have two choices depending on how "passion" is measured. If "passion" is an all-or-none phenomenon, then we can forget about trying to formulate a null hypothesis and instead test the binomial hypothesis that the probability p that a redhead is passionate is 80%. If "passion" can be measured on a seven-point scale and we define "passionate" as "passion" greater than or equal to 5, then our hypothesis becomes "the 20th percentile of redhead passion exceeds 5." As in the first example above, we could convert this to a "null hypothesis" by subtracting five from each observation. But the effort is unnecessary.

NEYMAN–PEARSON THEORY

Formulate your alternative hypotheses at the same time you set forth your principal hypothesis.

When the objective of our investigations is to arrive at some sort of conclusion, then we need to have not only a hypothesis in mind, but also one or more potential alternative hypotheses.

The cornerstone of modern hypothesis testing is the Neyman–Pearson Lemma. To get a feeling for the working of this lemma, suppose we are testing a new vaccine by administering it to half of our test subjects and giving a supposedly harmless placebo to each of the remainder. We proceed to follow these subjects over some fixed period and to note which subjects, if any, contract the disease that the new vaccine is said to offer protection against.

We know in advance that the vaccine is unlikely to offer complete protection; indeed, some individuals may actually come down with the disease as a result of taking the vaccine. Depending on the weather and other factors over which we have no control, our subjects, even those who received only placebo, may not contract the disease during the study period. All sorts of outcomes are possible.

The tests are being conducted in accordance with regulatory agency guidelines. From the regulatory agency's perspective, the principal hypothesis H is that the new vaccine offers no protection. Our alternative hypothesis A is that the new vaccine can cut the number of infected individuals in half. Our task before the start of the experiment is to decide which outcomes will rule in favor of the alternative hypothesis A and which in favor of the null hypothesis H.

The problem is that because of the variation inherent in the disease process, each and every one of the possible outcomes could occur regardless of which hypothesis is true. Of course, some outcomes are more likely if H is true (for example, 50 cases of pneumonia in the placebo group and

48 in the vaccine group), and others are more likely if the alternative hypothesis is true (for example, 38 cases of pneumonia in the placebo group and 20 in the vaccine group).

Following Neyman and Pearson, we order each of the possible outcomes in accordance with the ratio of its probability or likelihood when the alternative hypothesis is true to its probability when the principal hypothesis is true. When this likelihood ratio is large, we shall say the outcome rules in favor of the alternative hypothesis. Working downwards from the outcomes with the highest values, we continue to add outcomes to the *rejection* region of the test—so-called because these are the outcomes for which we would reject the primary hypothesis—until the total probability of the rejection region under the null hypothesis is equal to some predesignated *significance level*.

To see that we have done the best we can do, suppose we replace one of the outcomes we assigned to the rejection region with one we did not. The probability that this new outcome would occur if the primary hypothesis is true must be less than or equal to the probability that the outcome it replaced would occur if the primary hypothesis is true. Otherwise, we would exceed the significance level. Because of how we assigned outcome to the rejection region, the likelihood ratio of the new outcome is smaller than the likelihood ratio of the old outcome. Thus the probability the new outcome would occur if the alternative hypothesis is true must be less than or equal to the probability that the outcome it replaced would occur if the alternative hypothesis is true. That is, by swapping outcomes we have reduced the *power* of our test. By following the method of Neyman and Pearson and maximizing the likelihood ratio, we obtain the most powerful test at a given significance level.

To take advantage of Neyman and Pearson's finding, we need to have an alternative hypothesis or alternatives firmly in mind when we set up a test. Too often in published research, such alternative hypotheses remain unspecified or, worse, are specified only *after* the data are in hand. *We must specify our alternatives before we commence an analysis*, preferably at the same time we design our study.

Are our alternatives one-sided or two-sided? Are they ordered or unordered? The form of the alternative will determine the statistical procedures we use and the significance levels we obtain.

Decide beforehand whether you wish to test against a one-sided or a two-sided alternative.

One-Sided or Two-Sided

Suppose on examining the cancer registry in a hospital, we uncover the following data that we put in the form of a 2×2 contingency table.

	Survived	Died	Total
Men	9	1	10
Women	4	10	14
Total	13	11	24

The 9 denotes the number of males who survived, the 1 denotes the number of males who died, and so forth. The four marginal totals or marginals are 10, 14, 13, and 11. The total number of men in the study is 10, while 14 denotes the total number of women, and so forth.

The marginals in this table are fixed because, indisputably, there are 11 dead bodies among the 24 persons in the study and 14 women. Suppose that before completing the table, we lost the subject IDs so that we could no longer identify which subject belonged in which category. Imagine you are given two sets of 24 labels. The first set has 14 labels with the word "woman" and 10 labels with the word "man." The second set of labels has 11 labels with the word "dead" and 13 labels with the word "alive." Under the null hypothesis, you are allowed to distribute the labels to subjects independently of one another. One label from each of the two sets per subject, please.

There are a total of $\binom{24}{10}$ ways you could hand out the labels. $\binom{14}{10}\binom{10}{1}$ of the assignments result in tables that are as extreme as our original table (that is, in which 90% of the men survive) and $\binom{14}{11}\binom{10}{0}$ in tables that are more extreme (100% of the men survive). This is a very small fraction of the total, so we conclude that a difference in survival rates of the two sexes as extreme as the difference we observed in our original table is very unlikely to have occurred by chance alone. We reject the hypothesis that the survival rates for the two sexes are the same and accept the alternative hypothesis that, in this instance at least, males are more likely to profit from treatment (Table 2.1).

In the preceding example, we tested the hypothesis that survival rates do not depend on sex against the alternative that men diagnosed with cancer are likely to live longer than women similarly diagnosed. We rejected the null hypothesis because only a small fraction of the possible tables were as extreme as the one we observed initially. This is an example of a one-tailed test. But is it the correct test? Is this really the alternative hypothesis we would have proposed if we had not already seen the data? Wouldn't we have been just as likely to reject the null hypothesis that men

TABLE 2.1 Survial Rates of Men and Women[a]

	Survived	Died	Total
Men	10	0	10
Women	3	11	14
Total	13	11	24
	Survived	Died	Total
Men	8	2	10
Women	5	9	14
Total	13	11	24

[a] In terms of the Relative Survival Rates of the Two Sexes, the first of these tables is more extreme than our original table. The second is less extreme.

and women profit the same from treatment if we had observed a table of the following form?

	Survived	Died	Total
Men	0	10	10
Women	13	1	14
Total	13	11	24

Of course, we would! In determining the significance level in the present example, we must add together the total number of tables that lie in either of the two extremes or tails of the permutation distribution.

The critical values and significance levels are quite different for one-tailed and two-tailed tests; all too often, the wrong test has been employed in published work. McKinney et al. [1989] reviewed some 70 plus articles that appeared in six medical journals. In over half of these articles, Fisher's exact test was applied improperly. Either a one-tailed test had been used when a two-tailed test was called for or the authors of the paper simply hadn't bothered to state which test they had used.

Of course, unless you are submitting the results of your analysis to a regulatory agency, no one will know whether you originally intended a one-tailed test or a two-tailed test and subsequently changed your mind. No one will know whether your hypothesis was conceived before you started or only after you'd examined the data. All you have to do is lie. Just recognize that if you test an after-the-fact hypothesis without identifying it as such, you are guilty of scientific fraud.

When you design an experiment, decide at the same time whether you wish to test your hypothesis against a two-sided or a one-sided alternative.

A two-sided alternative dictates a two-tailed test; a one-sided alternative dictates a one-tailed test.

As an example, suppose we decide to do a follow-on study of the cancer registry to confirm our original finding that men diagnosed as having tumors live significantly longer than women similarly diagnosed. In this follow-on study, we have a one-sided alternative. Thus, we would analyze the results using a one-tailed test rather than the two-tailed test we applied in the original study.

Determine beforehand whether your alternative hypotheses are ordered or unordered.

Ordered or Unordered Alternative Hypotheses?

When testing qualities (number of germinating plants, crop weight, etc.) from k samples of plants taken from soils of different composition, it is often routine to use the F ratio of the analysis of variance. For contingency tables, many routinely use the chi-square test to determine if the differences among samples are significant. But the F-ratio and the chi-square are what are termed omnibus tests, designed to be sensitive to all possible alternatives. As such, they are not particularly sensitive to ordered alternatives such "as more fertilizer more growth" or "more aspirin faster relief of headache." Tests for such ordered responses at k distinct treatment levels should properly use the Pitman correlation described by Frank, Trzos, and Good [1978] when the data are measured on a metric scale (e.g., weight of the crop). Tests for ordered responses in $2 \times C$ contingency tables (e.g., number of germinating plants) should use the trend test described by Berger, Permutt, and Ivanova [1998]. We revisit this topic in more detail in the next chapter.

DEDUCTION AND INDUCTION

When we determine a p value as we did in the example above, we apply a set of algebraic methods and deductive logic to *deduce* the correct value. The deductive process is used to determine the appropriate size of resistor to use in an electric circuit, to determine the date of the next eclipse of the moon, and to establish the identity of the criminal (perhaps from the fact the dog did not bark on the night of the crime). Find the formula, plug in the values, turn the crank, and out pops the result (or it does for Sherlock Holmes,[4] at least).

When we assert that for a given population a percentage of samples will have a specific composition, this also is a deduction. But when we make an

---

[4] See "Silver Blaze" by A. Conan-Doyle, *Strand Magazine*, December 1892.

inductive generalization about a population based upon our analysis of a sample, we are on shakier ground. It is one thing to assert that if an observation comes from a normal distribution with mean zero, the probability is one-half that it is positive. It is quite another if, on observing that half the observations in the sample are positive, we assert that half of all the possible observations that might be drawn from that population will be positive also.

Newton's Law of gravitation provided an almost exact fit (apart from measurement error) to observed astronomical data for several centuries; consequently, there was general agreement that Newton's generalization from observation was an accurate description of the real world. Later, as improvements in astronomical measuring instruments extended the range of the observable universe, scientists realized that Newton's Law was only a generalization and not a property of the universe at all. Einstein's Theory of Relativity gives a much closer fit to the data, a fit that has not been contradicted by any observations in the century since its formulation. But this still does not mean that relativity provides us with a complete, correct, and comprehensive view of the universe.

In our research efforts, the only statements we can make with God-like certainty are of the form "our conclusions fit the data." The true nature of the real world is unknowable. We can speculate, but never conclude.

LOSSES

In our first advanced course in statistics, we read in the first chapter of Lehmann [1986] that the "optimal" statistical procedure would depend on the losses associated with the various possible decisions. But on day one of our venture into the real world of practical applications, we were taught to ignore this principle.

At that time, the only computationally feasible statistical procedures were based on losses that were proportional to the square of the difference between estimated and actual values. No matter that the losses really might be proportional to the absolute value of those differences, or the cube, or the maximum over a certain range. Our options were limited by our ability to compute.

Computer technology has made a series of major advances in the past half century. What required days or weeks to calculate 40 years ago takes only milliseconds today. We can now pay serious attention to this long neglected facet of decision theory: the losses associated with the varying types of decision.

Suppose we are investigating a new drug: We gather data, perform a statistical analysis, and draw a conclusion. If chance alone is at work yielding exceptional values and we opt in favor of the new drug, we've made

TABLE 2.2 Decision-Making Under Uncertainty

The Facts	Our Decision	
No difference.	No difference.	Drug is better. *Type I error:* Manufacturer wastes money developing ineffective drug.
Drug is better.	*Type II error:* Manufacturer misses opportunity for profit. Public denied access to effective treatment.	

TABLE 2.3 Decision-Making Under Uncertainty

The Facts	Fears et al.'s Decision	
Compound not a carcinogen.	Not a carcinogen.	Compound a carcinogen. *Type I error:* Manufacturer misses opportunity for profit. Public denied access to effective treatment.
Compound a carcinogen.	*Type II error:* Patients die; families suffer; Manufacturer sued.	

an error. We also make an error if we decide there is no difference and the new drug really is better. These decisions and the effects of making them are summarized in Table 2.2.

We distinguish the two types of error because they have the quite different implications described in Table 2.2. As a second example, Fears, Tarone, and Chu [1977] use permutation methods to assess several standard screens for carcinogenicity. As shown in Table 2.3, their Type I error, a false positive, consists of labeling a relatively innocuous compound as carcinogenic. Such an action means economic loss for the manufacturer and the denial to the public of the compound's benefits. Neither consequence is desirable. But a false negative, a Type II error, is much worse because it would mean exposing a large number of people to a potentially lethal compound.

What losses are associated with the decisions you will have to make? Specify them now before you begin.

DECISIONS

The hypothesis/alternative duality is inadequate in most real-life situations. Consider the pressing problems of global warming and depletion of the ozone layer. We could collect and analyze yet another set of data and

TABLE 2.4 Effect of Global Warming

The Facts	President's Decision on Emissions		
	Reduce emissions	Gather more data	Change unnecessary
No effect	Economy disrupted	Sampling cost	
Burning of fossil fuels responsible		Sampling cost Decline in quality of life (irreversible?)	Decline in quality of life (irreversible?)

then, just as is done today, make one of three possible decisions: reduce emissions, leave emission standards alone, or sit on our hands and wait for more data to come in. Each decision has consequences as shown in Table 2.4.

As noted at the beginning of this chapter, it's essential that we specify in advance the actions to be taken for each potential result. Always suspect are after-the-fact rationales that enable us to persist in a pattern of conduct despite evidence to the contrary. If no possible outcome of a study will be sufficient to change our mind, then perhaps we ought not undertake such a study in the first place.

Every research study involves multiple issues. Not only might we want to know whether a measurable, biologically (or medically, physically, or sociologically) significant effect takes place, but also what the size of the effect is and the extent to which the effect varies from instance to instance. We would also want to know what factors, if any, will modify the size of the effect or its duration.

We may not be able to address all these issues with a single data set. A preliminary experiment might tell us something about the possible existence of an effect, along with rough estimates of its size and variability. It is hoped that we will glean enough information to come up with doses, environmental conditions, and sample sizes to apply in collecting and evaluating the next data set. A list of possible decisions after the initial experiment includes "abandon this line of research," "modify the environment and gather more data," and "perform a large, tightly controlled, expensive set of trials." Associated with each decision is a set of potential gains and losses. Common sense dictates that we construct a table similar to Table 2.2 or 2.3 before we launch a study.

For example, in clinical trials of a drug we might begin with some animal experiments, then progress to Phase I clinical trials in which, with the emphasis on safety, we look for the maximum tolerable dose. Phase I trials generally involve only a small number of subjects and a one-time or short-term intervention. An extended period of several months may be used for follow-up purposes. If no adverse effects are observed, we might decide to go ahead with a further or Phase II set of trials in the clinic in which our objective is to determine the minimum effective dose. Obviously, if the minimum effective dose is greater than the maximum tolerable dose, or if some dangerous side effects are observed that we didn't observe in the first set of trials, we'll abandon the drug and go on to some other research project. But if the signs are favorable, then and only then will we go to a set of Phase III trials involving a large number of subjects observed over an extended time period. Then, and only then, will we hope to get the answers to all our research questions.

Before you begin, list all the consequences of a study and all the actions you might take. Persist only if you can add to existing knowledge.

TO LEARN MORE

For more thorough accounts of decision theory, the interested reader is directed to Berger [1986], Blyth [1970], Cox [1958], DeGroot [1970], and Lehmann [1986]. For an applied perspective, see Clemen [1991], Berry [1995], and Sox et al. [1988].

Over 300 references warning of the misuse of null hypothesis testing can be accessed online at the URL http://www.cnr.colostate.edu/~anderson/thompson1.html. Alas, the majority of these warnings are ill informed, stressing errors that will not arise if you proceed as we recommend and place the emphasis on the why, not the what, of statistical procedures. Use statistics as a guide to decision making rather than a mandate.

Neyman and Pearson [1933] first formulated the problem of hypothesis testing in terms of two types of error. Extensions and analyses of their approach are given by Lehmann [1986] and Mayo [1996]. For more work along the lines proposed here, see Selike, Bayarri, and Berger [2001].

Clarity in hypothesis formulation is essential; ambiguity can only yield controversy; see, for example, Kaplan [2001].

Chapter 3

Collecting Data

GIGO Garbage in, garbage out.
"Fancy statistical methods will not rescue garbage data."
Course notes of Raymond J. Carroll [2001].

THE VAST MAJORITY OF ERRORS IN STATISTICS—AND, not incidentally, in most human endeavors—arise from a reluctance (or even an inability) to plan. Some demon (or demonic manager) seems to be urging us to cross the street before we've had the opportunity to look both ways. Even on those rare occasions when we do design an experiment, we seem more obsessed with the mechanics than with the concepts that underlie it.

In this chapter we review the fundamental concepts of experimental design, the determination of sample size, the assumptions that underlie most statistical procedures, and the precautions necessary to ensure that they are satisfied and that the data you collect will be representative of the population as a whole. We do not intend to replace a text on experiment or survey design, but to supplement it, providing examples and solutions that are often neglected in courses on the subject.

PREPARATION

The first step in data collection is to have a clear, preferably written statement of your objectives. In accordance with Chapter 1, you will have defined the population or populations from which you intend to sample and have identified the characteristics of these populations you wish to investigate.

You developed one or more well-formulated hypotheses (the topic of Chapter 2) and have some idea of the risks you will incur should your analysis of the collected data prove to be erroneous. You will need to

Common Errors in Statistics (and How to Avoid Them), by Phillip I. Good and James W. Hardin.
ISBN 0-471-46068-0   Copyright © 2003 John Wiley & Sons, Inc.

decide what you wish to observe and measure and how you will go about observing it.

Good practice is to draft the analysis section of your final report based on the conclusions you would like to make. What information do you need to justify these conclusions? All such information must be collected.

The next section is devoted to the choice of measuring devices, followed by sections on determining sample size and preventive steps to ensure your samples will be analyzable by statistical methods.

MEASURING DEVICES

Know what you want to measure. Collect exact values whenever possible.

Know what you want to measure. Will you measure an endpoint such as death or measure a surrogate such as the presence of HIV antibodies? The regression slope describing the change in systolic blood pressure (in mm Hg) per 100 mg of calcium intake is strongly influenced by the approach used for assessing the amount of calcium consumed (Cappuccio et al., 1995). The association is small and only marginally significant with diet histories (slope −0.01 (−0.003 to −0.016)) but large and highly significant when food frequency questionnaires are used (−0.15 (−0.11 to −0.19)). With studies using 24-hour recall, an intermediate result emerges (−0.06 (−0.09 to −0.03)). Diet histories assess patterns of usual intake over long periods of time and require an extensive interview with a nutritionist, whereas 24-hour recall, and food frequency questionnaires are simpler methods that reflect current consumption (Block, 1982).

Before we initiate data collection, we must have a firm idea of what we will measure.

A second fundamental principle is also applicable to both experiments and surveys: Collect exact values whenever possible. Worry about grouping them in interval or discrete categories later.

A long-term study of buying patterns in New South Wales illustrates some of the problems caused by grouping prematurely. At the beginning of the study, the decision was made to group the incomes of survey subjects into categories, under $20,000, $20,000 to $30,000, and so forth. Six years of steady inflation later, the organizers of the study realized that all the categories had to be adjusted. An income of $21,000 at the start of the study would only purchase $18,000 worth of goods and housing at the end. The problem was that those surveyed toward the end had filled out forms with exactly the same income categories. Had income been tabulated to the nearest dollar, it would have been easy to correct for increases in the cost of living and convert all responses to the same scale.

But the study designers hadn't considered these issues. A precise and costly survey was now a matter of guesswork.

You can always group your results (and modify your groupings) after a study is completed. If after-the-fact grouping is a possibility, your design should state how the grouping will be determined; otherwise there will be the suspicion that you chose the grouping to obtain desired results.

Experiments

Measuring devices differ widely both in what they measure and in the precision with which they measure it. As noted in the next section of this chapter, the greater the precision with which measurements are made, the smaller the sample size required to reduce both Type I and Type II errors below specific levels.

Before you rush out and purchase the most expensive and precise measuring instruments on the market, consider that the total cost C of an experimental procedure is $S + nc$, where n is the sample size and c is the cost per unit sampled.

The startup cost S includes the cost of the measuring device. c is made up of the cost of supplies and personnel costs. The latter includes not only the time spent on individual measurements but also the time spent in preparing and calibrating the instrument for use.

Less obvious factors in the selection of a measuring instrument include impact on the subject, reliability (personnel costs continue even when an instrument is down), and reusability in future trials. For example, one of the advantages of the latest technology for blood analysis is that less blood needs to be drawn from patients. Less blood means happier subjects, fewer withdrawals, and a smaller initial sample size.

Surveys

While no scientist would dream of performing an experiment without first mastering all the techniques involved, an amazing number will blunder into the execution of large-scale and costly surveys without a preliminary study of all the collateral issues a survey entails.

We know of one institute that mailed out some 20,000 questionnaires (didn't the post office just raise its rates again?) before discovering that half the addresses were in error and that the vast majority of the remainder were being discarded unopened before prospective participants had even read the "sales pitch."

Fortunately, there are texts such as Bly [1990, 1996] that will tell you how to word a "sales pitch" and the optimal colors and graphics to use along with the wording. They will tell you what "hooks" to use on the envelope to ensure attention to the contents and what premiums to offer to increase participation.

There are other textbooks such as Converse and Presser [1986], Fowler and Fowler [1995], and Schroeder [1987] to assist you in wording questionnaires and in pretesting questions for ambiguity before you begin. We have only two paragraphs of caution to offer:

1. Be sure your questions don't reveal the purpose of your study; otherwise, respondents shape their answers to what they perceive to be your needs. Contrast "how do you feel about compulsory pregnancy?" with "how do you feel about abortions?"

2. With populations ever more heterogeneous, questions that work with some ethnic groups may repulse others (see, for example, Choi [2000]).

Recommended are web-based surveys with initial solicitation by mail (letter or post card) and email. Not only are both costs and time to completion cut dramatically, but also the proportion of missing data and incomplete forms is substantially reduced. Moreover, web-based surveys are easier to monitor, and forms may be modified on the fly. Web-based entry also offers the possibility of displaying the individual's prior responses during follow-up surveys.

Three other precautions can help ensure the success of your survey:

1. Award premiums only for fully completed forms.

2. Continuously tabulate and monitor submissions; don't wait to be surprised.

3. A quarterly newsletter sent to participants will substantially increase retention (and help you keep track of address changes).

DETERMINING SAMPLE SIZE

Determining optimal sample size is simplicity itself once we specify all of the following:

- Desired power and significance level.
- Distributions of the observables.
- Statistical test(s) that will be employed.
- Anticipated losses due to nonresponders, noncompliant participants, and dropouts.

Power and Significance Level

Understand the relationships among sample size, significance level, power, and precision of the measuring instruments.

Sample size must be determined for each experiment; there is no universally correct value (Table 3.1). Increase the precision (and hold all other parameters fixed) and we can decrease the required number of observations.

TABLE 3.1 Ingredients in a Sample Size Calculation

Type I error (α)	Probability of falsely rejecting the hypothesis when it is true.
Type II error ($1 - \beta[A]$)	Probability of falsely accepting the hypothesis when an alternative hypothesis A is true. Depends on the alternative A.
Power $= \beta[A]$	Probability of correctly rejecting the hypothesis when an alternative hypothesis A is true. Depends on the alternative A.
Distribution functions	$F[(x - \mu)\sigma]$, e.g., normal distribution.
Location parameters	For both hypothesis and alternative hypothesis: μ_1, μ_2.
Scale parameters	For both hypothesis and alternative hypothesis: σ_1, σ_2.
Sample sizes	May be different for different groups in an experiment with more than one group

Permit a greater number of Type I or Type II errors (and hold all other parameters fixed) and we can decrease the required number of observations.

Explicit formula for power and significance level are available when the underlying observations are binomial, the results of a counting or Poisson process, or normally distributed. Several off-the-shelf computer programs including nQuery Advisor™, Pass 2000™, and StatXact™ are available to do the calculations for us.

To use these programs, we need to have some idea of the location (mean) and scale parameter (variance) of the distribution both when the primary hypothesis is true and when an alternative hypothesis is true. Since there may well be an infinity of alternatives in which we are interested, power calculations should be based on the worst-case or boundary value. For example, if we are testing a binomial hypothesis $p = 1/2$ against the alternatives $p \leq 2/3$, we would assume that $p = 2/3$.

If the data do not come from one of the preceding distributions, then we might use a bootstrap to estimate the power and significance level.

In preliminary trials of a new device, the following test results were observed: 7.0 in 11 out of 12 cases and 3.3 in 1 out of 12 cases. Industry guidelines specified that any population with a mean test result greater than 5 would be acceptable. A worst-case or boundary-value scenario would include one in which the test result was 7.0 3/7th of the time, 3.3 3/7th of the time, and 4.1 1/7th of the time.

The statistical procedure required us to reject if the sample mean of the test results were less than 6. To determine the probability of this event for various sample sizes, we took repeated samples with replacement from the two sets of test results. Some bootstrap samples consisted of all 7's, whereas some, taken from the worst-case distribution, consisted only of

TABLE 3.2 Power Estimates

Sample Size	Test Mean < 6	
	α	Power
3	0.23	0.84
4	0.04	0.80
5	0.06	0.89

3.3's. Most were a mixture. Table 3.2 illustrates the results; for example, in our trials, 23% of the bootstrap samples of size 3 from our starting sample of test results had medians less than 6. If, instead, we drew our bootstrap samples from the hypothetical "worst-case" population, then 84% had medians less than 6.

If you want to try your hand at duplicating these results, simply take the test values in the proportions observed, stick them in a hat, draw out bootstrap samples with replacement several hundred times, compute the sample means, and record the results. Or you could use the Stata™ bootstrap procedure as we did.[1]

Prepare for Missing Data

The relative ease with which a program like Stata or StatXact can produce a sample size may blind us to the fact that the number of subjects with which we begin a study may bear little or no relation to the number with which we conclude it.

A midsummer hailstorm, an early frost, or an insect infestation can lay waste to all or part of an agricultural experiment. In the National Institute of Aging's first years of existence, a virus wiped out the entire primate colony destroying a multitude of experiments in progress.

Large-scale clinical trials and surveys have a further burden, namely, the subjects themselves. Potential subjects can and do refuse to participate. (Don't forget to budget for a follow-up study, bound to be expensive, of responders versus nonresponders.) Worse, they agree to participate initially, then drop out at the last minute (see Figure 3.1).

They move without a forwarding address before a scheduled follow-up. Or simply don't bother to show up for an appointment. We lost 30% of the patients in the follow-ups to a lifesaving cardiac procedure. (We can't imagine not going in to see our surgeon, but then we guess we're not typical.)

The key to a successful research program is to plan for such dropouts in advance and to start the trials with some multiple of the number required to achieve a given power and significance level.

---

[1] Chapters 4–7 have more information on the use of the bootstrap and its limitations.

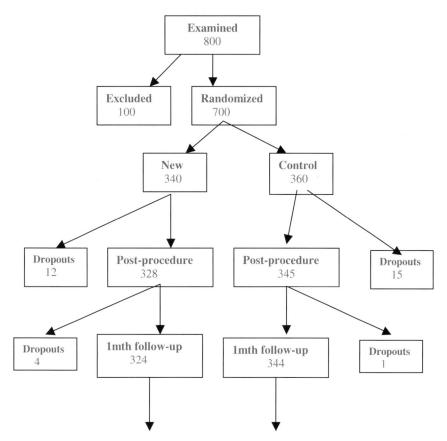

FIGURE 3.1 **A Typical Clinical Trial.** Dropouts and noncompliant patients occur at every stage. Reprinted from the *Manager's Guide to Design and Conduct of Clinical Trials* with the permission of John Wiley & Sons, Inc.

Nonresponders

An analysis of those who did not respond to a survey or a treatment can sometimes be as informative as, or more informative than, the survey itself. See, for example, Mangel and Samaniego [1984] as well as the sections on the Behrens–Fisher problem and on the premature drawing of conclusions in Chapter 5. Be sure to incorporate provisions for sampling nonresponders in your sample design and in your budget.

Sample from the Right Population

Be sure you are sampling from the population as a whole rather than from an unrepresentative subset of the population. The most famous blunder along these lines was basing the forecast of Landon over Roosevelt in the 1936 U.S. presidential election on a telephone survey; those who owned a telephone and responded to the survey favored Landon; those who voted did not. An economic study may be flawed because we have overlooked the homeless,[2] and an astrophysical study may be flawed because of over-looking galaxies whose central surface brightness was very low.[3]

FUNDAMENTAL ASSUMPTIONS

Most statistical procedures rely on two fundamental assumptions: that the observations are independent of one another and that they are identically distributed. If your methods of collection fail to honor these assumptions, then your analysis must fail also.

Independent Observations

To ensure the independence of responses in a return-by-mail or return-by-web survey, no more than one form per household should be accepted. If a comparison of the responses within a household is desired, then the members of the household should be interviewed separately, outside of each other's hearing, and with no opportunity to discuss the survey in between. People care what other people think and when asked about an emotionally charged topic may or may not tell the truth. In fact, they are unlikely to tell the truth if they feel that others may overhear or somehow learn of their responses.

To ensure independence of the observations in an experiment, determine in advance what constitutes the *experimental unit*.

In the majority of cases, the unit is obvious: One planet means one position in space, one container of gas means one volume and pressure to be recorded, and one runner on one fixed race course means one elapsed time.

In a clinical trial, each individual patient corresponds to a single set of observations or does she? Suppose we are testing the effects of a topical ointment on pinkeye. Is each eye a separate experimental unit, or each patient?

It is common in toxicology to examine a large number of slides. But regardless of how many are examined in the search for mutagenic and toxic effects, if all slides come from a single treated animal, then the total size of the sample is one.

We may be concerned with the possible effects a new drug might have on a pregnant woman and, as critically, on her children. In our preliminary tests, we'll be working with mice. Is each fetus in the litter a separate experimental unit, or each mother?

---

[2] *City of New York v. Dept of Commerce*, 822 F. Supp. 906 (E.D.N.Y., 1993).
[3] Bothun [1998, p. 249].

If the mother is the one treated with the drug, then the mother is the experimental unit, not the fetus. A litter of six or seven corresponds only to a sample of size one.

As for the topical ointment, while more precise results might be obtained by treating only one eye with the new ointment and recording the subsequent difference in appearance between the treated and untreated eyes, each patient still yields only one observation, not two.

Identically Distributed Observations

If you change measuring instruments during a study or change observers, then you will have introduced an additional source of variation and the resulting observations will not be identically distributed.

The same problems will arise if you discover during the course of a study (as is often the case) that a precise measuring instrument is no longer calibrated and readings have drifted. To forestall this, any measuring instrument should have been exposed to an extensive burn-in before the start of a set of experiments and should be recalibrated as frequently as the results of the burn-in or pre-study period dictate.

Similarly, one doesn't just mail out several thousand copies of a survey before performing an initial pilot study to weed out or correct ambiguous and misleading questions.

The following groups are unlikely to yield identically distributed observations: the first to respond to a survey, those who only respond after been offered an inducement, and nonresponders.

EXPERIMENTAL DESIGN

Statisticians have found three ways for coping with individual-to-individual and observer-to-observer variation:

1. *Controlling.* Making the environment for the study—the subjects, the manner in which the treatment is administered, the manner in which the observations are obtained, the apparatus used to make the measurements, and the criteria for interpretation—as uniform and homogeneous as possible.

2. *Blocking.* A clinician might stratify the population into subgroups based on such factors as age, sex, race, and the severity of the condition and restricting comparisons to individuals who belong to the same subgroup. An agronomist would want to stratify on the basis of soil composition and environment.

3. *Randomizing.* Randomly assigning patients to treatment within each subgroup so that the innumerable factors that can neither be controlled nor observed directly are as likely to influence the outcome of one treatment as another.

Steps 1 and 2 are trickier than they appear at first glance. Do the phenomena under investigation depend upon the time of day as with body temperature and the incidence of mitosis? Do they depend upon the day of the week as with retail sales and the daily mail? Will the observations be affected by the sex of the observer? Primates (including you) and hunters (tigers, mountain lions, domestic cats, dogs, wolves, and so on) can readily detect the observer's sex.[4]

Blocking may be mandatory because even a randomly selected sample may not be representative of the population as a whole. For example, if a minority comprises less than 10% of a population, then a jury of 12 persons selected at random from that population will fail to contain a single member of that minority at least 28% of the time.

Groups to be compared may differ in other important ways even before any intervention is applied. These baseline imbalances cannot be attributed to the interventions, but they can interfere with and overwhelm the comparison of the interventions.

One good after-the-fact solution is to break the sample itself into strata (men, women, Hispanics) and to extrapolate separately from each stratum to the corresponding subpopulation from which the stratum is drawn.

The size of the sample we take from each block or strata need not, and in some instances should not, reflect the block's proportion in the population. The latter exception arises when we wish to obtain separate estimates for each subpopulation. For example, suppose we are studying the health of Marine recruits and wish to obtain separate estimates for male and female Marines as well as for Marines as a group. If we want to establish the incidence of a relatively rare disease, we will need to oversample female recruits to ensure that we obtain a sufficiently large number. To obtain a rate R for *all* Marines, we would then take the weighted average $p_F R_F + p_M R_M$ of the separate rates for each gender, where the proportions p_M and p_F are those of males and females in the *entire* population of Marine recruits.

FOUR GUIDELINES

In the next few sections on experimental design, we may well be preaching to the choir, for which we apologize. But there is no principle of experimental design, however obvious, and however intuitive, that someone will not argue can be ignored in his or her special situation:

> • Physicians feel they should be allowed to select the treatment that will best affect their patient's condition (but who is to know in advance what this treatment is?).

---

[4] The hair follicles of redheads—genuine, not dyed—are known to secrete a prostaglandin similar to an insect pheromone.

- Scientists eject us from their laboratories when we suggest that only the animal caretakers be permitted to know which cage houses the control animals.
- Engineers at a firm that specializes in refurbishing medical devices objected when Dr. Good suggested that they purchase and test some new equipment for use as controls. "But that would cost a fortune."

The statistician's lot is not a happy one. The opposite sex ignores us because we are boring,[5] and managers hate us because all our suggestions seem to require an increase in the budget. But controls will save money in the end. Blinding is essential if our results are to have credence, and care in treatment allocation is mandatory if we are to avoid bias.

Randomize

Permitting treatment allocation by either experimenter or subject will introduce bias.

Controls

To guard against the unexpected, as many or more patients should be assigned to the control regimen as are assigned to the experimental one. This sounds expensive and is. But shit happens. You get the flu. You get a headache or the runs. You have a series of colds that blend one into the other until you can't remember the last time you were well. So you blame your silicone implants. Or, if you are part of a clinical trial, you stop taking the drug. It's in these and similar instances that experimenters are grateful they've included controls. This is because when the data are examined, experimenters learn that as many of the control patients came down with the flu as those who were on the active drug, and they also learn that those women without implants had exactly the same incidence of colds and headaches as those who had implants.

Reflect on the consequences of not using controls. The first modern silicone implants (Dow Corning's Silastic mammary prosthesis) were placed in 1962. In 1984, a jury awarded $2 million to a recipient who complained of problems resulting from the implants. Award after award followed, the largest being more than $7 million. A set of controlled randomized trials was finally begun in 1994. The verdict: Silicon implants have no adverse effects on recipients. Tell this to the stockholders of bankrupt Dow Corning.

Use positive controls.

---

[5] Dr. Good told his wife he was an author—it was the only way he could lure someone that attractive to his side. Dr. Hardin is still searching for an explanation for his own good fortune.

There is no point in conducting an experiment if you already know the answer.[6] The use of a positive control is always to be preferred. A new anti-inflammatory should be tested against aspirin or ibuprofen. And there can be no justification whatever for the use of placebo in the treatment of a life-threatening disease (Barbui et al., 2000; Djulbegovic et al., 2000).

Blind Observers

Observers should be blinded to the treatment allocation.

Patients often feel better solely because they think they ought to feel better. A drug may not be effective if the patient is aware it is the old or less-favored remedy. Nor is the patient likely to keep taking a drug on schedule if he or she feels the pill contains nothing of value. She is also less likely to report any improvement in her condition, if she feels the doctor has done nothing for her. Vice versa, if a patient is informed she has the new treatment, she may think it necessary to "please the doctor" by reporting some diminishment in symptoms. These sorts of behavioral phenomena are precisely the reason why clinical trials must include a control.

A double-blind study in which neither the physician nor the patient knows which treatment is received is preferable to a single-blind study in which only the patient is kept in the dark (Ederer, 1975; Chalmers et al., 1983; Vickers et al., 1997).

Even if a physician has no strong feelings one way or the other concerning a treatment, she may tend to be less conscientious about examining patients she knows belong to the control group. She may have other unconscious feelings that influence her work with the patients, and she may have feelings about the patients themselves. Exactly the same caveats apply in work with animals and plants; units subjected to the existing, less-important treatment may be handled more carelessly and may be less thoroughly examined.

We recommend you employ two or even three individuals, one to administer the intervention, one to examine the experimental subject, and a third to observe and inspect collateral readings such as angiograms, laboratory findings, and x-rays that might reveal the treatment.

Conceal Treatment Allocation

Without allocation concealment, selection bias can invalidate study results (Schultz, 1995; Berger and Exner, 1999). If an experimenter could predict the next treatment to be assigned, he might exercise an unconscious bias in the treatment of that patient; he might even defer enrollment of a

---

[6] The exception being to satisfy a regulatory requirement.

patient he considers less desirable. In short, randomization alone, without allocation concealment, is insufficient to eliminate selection bias and ensure the internal validity of randomized clinical trials.

Lovell et al. [2000] describe a study in which four patients were randomized to the wrong stratum; in two cases, the treatment received was reversed. For an excruciatingly (and embarrassingly) detailed analysis of this experiment by an FDA regulator, see http://www.fda.gov/cber/review/etanimm052799r2.pdf.

Vance Berger and Costas Christophi offer the following guidelines for treatment allocation:

- Generate the allocation sequence in advance of screening any patients.
- Conceal the sequence from the experimenters.
- Require the experimenter to enroll all eligible subjects in the order in which they are screened.
- Verify that the subject actually received the assigned treatment.
- Conceal the proportions that have already been allocated (Schultz, 1996).
- Conceal treatment codes until all patients have been randomized and the database is locked.
- Do not permit enrollment discretion when randomization may be triggered by some earlier response pattern.

Blocked Randomization, Restricted Randomization, and Adaptive Designs

All the above caveats apply to these procedures as well. The use of an advanced statistical technique does not absolve its users from the need to exercise common sense. Observers must be kept blinded to the treatment received.

TO LEARN MORE

Good [2002] provides a series of anecdotes concerning the mythical Bumbling Pharmaceutical and Device Company that amply illustrate the results of inadequate planning. See also Andersen [1990] and Elwood [1998].

Definitions and a further discussion of the interrelation among power and significance level may be found in Lehmann [1986], Casella and Berger [1990], and Good [2001]. You'll also find discussions of optimal statistical procedures and their assumptions.

Shuster [1993] offers sample size guidelines for clinical trials. A detailed analysis of bootstrap methodology is provided in Chapters 3 and 7.

For further insight into the principles of experimental design, light on math and complex formulas but rich in insight, study the lessons of the masters: Fisher [1925, 1935] and Neyman [1952]. If formulas are what you desire, see Thompson and Seber [1996], Rosenbaum [2002], Jennison and Turnbull [1999], and Toutenburg [2002].

Among the many excellent texts on survey design are Fink and Kosecoff [1988], Rea, Parker, and Shrader [1997], and Cochran [1977]. For tips on formulating survey questions, see Converse and Presser [1986], Fowler and Fowler [1995], and Schroeder [1987]. For tips on improving the response rate, see Bly [1990, 1996].

Part II
HYPOTHESIS TESTING AND ESTIMATION

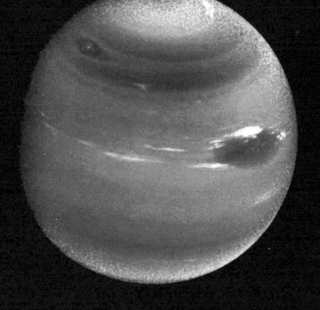

Chapter 4
Estimation

ACCURATE, RELIABLE ESTIMATES ARE ESSENTIAL TO EFFECTIVE DECISION-
MAKING. In this chapter, we review preventive measures and list the properties to look for in an estimation method. Several robust semiparametric
estimators are considered along with one method of interval estimation,
the bootstrap.

PREVENTION

The vast majority of errors in estimation stem from a failure to measure
what one wanted to measure or what one thought one was measuring.
Misleading definitions, inaccurate measurements, errors in recording and
transcription, and confounding variables plague results.

To forestall such errors, review your data collection protocols and procedure manuals before you begin, run several preliminary trials, record
potential confounding variables, monitor data collection, and review the
data as they are collected.

DESIRABLE AND NOT-SO-DESIRABLE ESTIMATORS

"The method of maximum likelihood is, by far, the most popular technique for deriving estimators" Casella and Berger [1990, p. 289]. The
proper starting point for the selection of the "best" method of estimation
is with the objectives of our study: What is the purpose of our estimate?
If our estimate is θ^* and the actual value of the unknown parameter is θ,
what losses will we be subject to? It is difficult to understand the popular-

Common Errors in Statistics (and How to Avoid Them), by Phillip I. Good and James W. Hardin.
ISBN 0-471-46068-0   Copyright © 2003 John Wiley & Sons, Inc.

ity of the method of maximum likelihood and other estimation procedures that do not take these losses into consideration.

The majority of losses will be monotone nondecreasing in nature; that is, the further apart the estimate θ^* and the true value θ, the larger our losses are likely to be. Typical forms of the loss function are the absolute deviation $|\theta^* - \theta|$, the square deviation $(\theta^* - \theta)^2$, and the jump—that is, no loss if $|\theta^* - \theta| < \delta$, and a big loss otherwise. Or the loss function may resemble the square deviation but take the form of a step function increasing in discrete increments.

Desirable estimators share the following properties: impartial, consistent, efficient, robust, and minimum loss.

Impartiality

Estimation methods should be impartial. Decisions should not depend on the accidental and quite irrelevant labeling of the samples. Nor should decisions depend on the units in which the measurements are made.

Suppose we have collected data from two samples with the object of estimating the difference in location of the two populations involved. Suppose further that the first sample includes the values a, b, c, d, and e, the second sample includes the values f, g, h, i, j, k, and our estimate of the difference is θ^*. If the observations are completely reversed—that is, if the first sample includes the values f, g, h, i, j, k and the second sample the values a, b, c, d, and e—our estimation procedure should declare the difference to be $-\theta^*$.

The units we use in our observations should not affect the resulting estimates. We should be able to take a set of measurements in feet, convert to inches, make our estimate, convert back to feet, and get absolutely the same result as if we'd worked in feet throughout. Similarly, where we locate the zero point of our scale should not affect the conclusions.

Finally, if our observations are independent of the time of day, the season, and the day on which they were recorded (facts that ought to be verified before proceeding further), then our estimators should be independent of the order in which the observations were collected.

Consistency

Estimators should be *consistent*; that is, the larger the sample, the greater the probability the resultant estimate will be close to the true population value.

Efficient

One consistent estimator certainly is to be preferred to another if the first consistent estimator can provide the same degree of accuracy with fewer observations. To simplify comparisons, most statisticians focus on the *asymptotic relative efficiency* (ARE), defined as the limit with increasing sample size of the ratio of the number of observations required for each of two consistent statistical procedures to achieve the same degree of accuracy.

Robust

Estimators that are perfectly satisfactory for use with symmetric normally distributed populations may not be as desirable when the data come from nonsymmetric or heavy-tailed populations, or when there is a substantial risk of contamination with extreme values.

When estimating measures of central location, one way to create a more robust estimator is to trim the sample of its minimum and maximum values (the procedure used when judging ice-skating or gymnastics). As information is thrown away, trimmed estimators are less efficient.

In many instances, LAD (least absolute deviation) estimators are more robust than their LS (least square) counterparts.[1] This finding is in line with our discussion of the *F* statistic in the preceding chapter.

Many *semiparametric estimators* are not only robust but provide for high ARE with respect to their parametric counterparts.

As an example of a semi-parametric estimator, suppose the $\{X_i\}$ are independent identically distributed (i.i.d.) observations with distribution $\Pr\{X_i \le x\} = F[y - \Delta]$ and we want to estimate the location parameter Δ without having to specify the form of the distribution *F*. If *F* is normal and the loss function is proportional to the square of the estimation error, then the arithmetic mean is optimal for estimating Δ. Suppose, on the other hand, that *F* is symmetric but more likely to include very large or very small values than a normal distribution. Whether the loss function is proportional to the absolute value or the square of the estimation error, the median, a semiparametric estimator, is to be preferred. The median has an ARE relative to the mean that ranges from 0.64 (if the observations really do come from a normal distribution) to values well in excess of 1 for distributions with higher proportions of very large and very small values (Lehmann, 1998, p. 242). Still, if the unknown distribution is "almost" normal, the mean would be far preferable.

If we are uncertain whether or not *F* is symmetric, then our best choice is the Hodges–Lehmann estimator defined as the median of the pairwise averages

$$\hat{\Delta} = \text{median}_{i \le j} (X_j + X_i)/2.$$

---

[1] See, for example, Yoo [2001].

Its ARE relative to the mean is 0.97 when F is a normal distribution (Lehmann, 1998, p. 246). With little to lose with respect to the mean if F is near normal, and much to gain if F is not, the Hodges–Lehmann estimator is recommended.

Suppose $\{X_i\}$ and $\{Y_j\}$ are i.i.d. with distributions $\Pr\{ X_i \leq x\} = F[x]$ and $\Pr \{ Y_j \leq y\} = F[y - \Delta]$ and we want to estimate the shift parameter Δ without having to specify the form of the distribution F. For a normal distribution F, the optimal estimator with least-square losses is

$$\overline{\Delta} = \frac{1}{mn}\sum_i \sum_j (Y_j - X_i) = \overline{Y} - \overline{X},$$

the arithmetic average of the mn differences $Y_j - X_i$. Means are highly dependent on extreme values; a more robust estimator is given by

$$\hat{\Delta} = \text{median}_{ij}(X_j - X_i).$$

Minimum Loss

The value taken by an estimate, its accuracy (that is, the degree to which it comes close to the true value of the estimated parameter), and the associated losses will vary from sample to sample. *A minimum loss estimator* is one that minimizes the losses when the losses are averaged over the set of all possible samples. Thus its form depends upon all of the following: the loss function, the population from which the sample is drawn, and the population characteristic that is being estimated. An estimate that is optimal in one situation may only exacerbate losses in another.

Minimum loss estimators in the case of least-square losses are widely and well documented for a wide variety of cases. Linear regression with an LAD loss function is discussed in Chapter 9.

Mini–Max Estimators

It's easy to envision situations in which we are less concerned with the average loss than with the maximum possible loss we may incur by using a particular estimation procedure. An estimate that minimizes the maximum possible loss is termed a mini–max estimator. Alas, few off-the-shelf mini–max solutions are available for practical cases, but see Pilz [1991] and Pinelis [1988].

Other Estimation Criteria

The expected value of an *unbiased* estimator is the population characteristic being estimated. Thus, unbiased estimators are also consistent estimators.

Minimum variance estimators provide relatively consistent results from sample to sample. While minimum variance is desirable, it may be of practical value only if the estimator is also *unbiased*. For example, 6 is a minimum variance estimator, but offers few other advantages.

Plug-in estimators, in which one substitutes the sample statistic for the population statistic, the sample mean for the population mean, or the sample's 20th percentile for the population's 20th percentile, are consistent, but they are not always unbiased or minimum loss.

Always choose an estimator that will minimize losses.

Myth of Maximum Likelihood

The popularity of the maximum likelihood estimator is hard to comprehend. This estimator may be completely unrelated to the loss function and has as its sole justification that it corresponds to that value of the parameter that makes the observations most probable—provided, that is, they are drawn from a specific predetermined distribution. The observations might have resulted from a thousand other a priori possibilities.

A common and lamentable fallacy is that the maximum likelihood estimator has many desirable properties—that it is unbiased and minimizes the mean-squared error. But this is true only for the maximum likelihood estimator of the mean of a normal distribution.[2]

Statistics instructors would be well advised to avoid introducing maximum likelihood estimation and to focus instead on methods for obtaining minimum loss estimators for a wide variety of loss functions.

INTERVAL ESTIMATES

Point estimates are seldom satisfactory in and of themselves. First, if the observations are continuous, the probability is zero that a point estimate will be correct and equal the estimated parameter. Second, we still require some estimate of the precision of the point estimate.

In this section, we consider one form of *interval estimate* derived from bootstrap measures of precision. A second form, derived from tests of hypotheses, will be considered in the next chapter.

Nonparametric Bootstrap

The bootstrap can help us obtain an interval estimate for any aspect of a distribution—a median, a variance, a percentile, or a correlation coefficient—*if* the observations are independent and all come from distributions

---

[2] It is also true in some cases for very large samples. How large the sample must be in each case will depend both upon the parameter being estimated and upon the distribution from which the observations are drawn.

with the same value of the parameter to be estimated. This interval provides us with an estimate of the precision of the corresponding point estimate.

We resample with replacement repeatedly from the original sample, 1000 times or so, computing the sample statistic for each bootstrap sample.

For example, here are the heights of a group of 22 adolescents, measured in centimeters and ordered from shortest to tallest.

137.0 138.5 140.0 141.0 142.0 143.5 145.0 147.0 148.5 150.0 153.0 154.0 155.0 156.5 157.0 158.0 158.5 159.0 160.5 161.0 162.0 167.5

The median height lies somewhere between 153 and 154 cm. If we want to extend this result to the population, we need an estimate of the precision of this average.

Our first bootstrap sample, arranged in increasing order of magnitude for ease in reading, might look like this:

138.5 138.5 140.0 141.0 141.0 143.5 145.0 147.0 148.5 150.0 153.0 154.0 155.0 156.5 157.0 158.5 159.0 159.0 159.0 160.5 161.0 162.0

Several of the values have been repeated; this is not surprising because we are sampling with replacement, treating the original sample as a stand-in for the much larger population from which the original sample was drawn. The minimum of this bootstrap sample is 138.5, higher than that of the original sample; the maximum at 162.0 is less than the original, while the median remains unchanged at 153.5.

137.0 138.5 138.5 141.0 141.0 142.0 143.5 145.0 145.0 147.0 148.5 148.5 150.0 150.0 153.0 155.0 158.0 158.5 160.5 160.5 161.0 167.5

In this second bootstrap sample, again we find repeated values; this time the minimum, maximum, and median are 137.0, 167.5, and 148.5, respectively.

The medians of 50 bootstrapped samples drawn from our sample ranged between 142.25 and 158.25, with a median of 152.75 (see Figure 4.1). These numbers provide an insight into what might have been had we sampled repeatedly from the original population.

FIGURE 4.1 Scatterplot of 50 Bootstrap Medians Derived from a Sample of Heights.

We can improve on the interval estimate {142.25, 158.25} if we are willing to accept a small probability that the interval will fail to include the true value of the population median. We will take several hundred bootstrap samples instead of a mere 50, and we will use the 5th and 95th percentiles of the resulting bootstrap distribution to establish the boundaries of a 90% confidence interval.

This method might be used equally well to obtain an interval estimate for any other population attribute: the mean and variance, the 5th percentile or the 25th, and the interquartile range. When several observations are made simultaneously on each subject, the bootstrap can be used to estimate covariances and correlations among the variables. The bootstrap is particularly valuable when trying to obtain an interval estimate for a ratio or for the mean and variance of a nonsymmetric distribution.

Unfortunately, such intervals have two deficiencies:

1. **They are biased; that is, they are more likely to contain certain false values of the parameter being estimated than the true one (Efron, 1987).**

2. **They are wider and less efficient than they could be (Efron, 1987).**

Two methods have been proposed to correct these deficiencies; let us consider each in turn.

The first is the Hall–Wilson [Hall and Wilson, 1991] corrections in which the bootstrap estimate is Studentized. For the one-sample case, we want an interval estimate based on the distribution of $(\hat{\theta}_b - \hat{\theta})/s_b$, where $\hat{\theta}$ and $\hat{\theta}_b$ are the estimates of the unknown parameter based on the original and bootstrap sample, respectively, and s_b denotes the standard deviation of the bootstrap sample. An estimate $\hat{\sigma}$ of the population variance is required to transform the resultant interval into one about θ (see Carpenter and Bithell [2000]).

For the two-sample case, we want a confidence interval based on the distribution of

$$\frac{\left(\hat{\theta}_{nb} - \hat{\theta}_{mb}\right)}{\sqrt{\dfrac{(n-1)s_{nb}^2 + (m-1)s_{mb}^2}{n+m-2}\left(1/n + 1/m\right)}},$$

where n, m, and s_{nb}, s_{mb} denote the sample sizes and standard deviations, respectively, of the bootstrap samples. Applying the Hall–Wilson corrections, we obtain narrower interval estimates that are more likely to contain the true value of the unknown parameter.

The bias-corrected and accelerated BC_a interval due to Efron and Tibshirani [1986] also represents a substantial improvement, though for samples under size 30, the interval is still suspect. The idea behind these intervals comes from the observation that percentile bootstrap intervals are most accurate when the estimate is symmetrically distributed about the true value of the parameter and the tails of the estimate's distribution drop off rapidly to zero. The symmetric, bell-shaped normal distribution depicted in Figure 7.1 represents this ideal.

Suppose θ is the parameter we are trying to estimate, $\hat{\theta}$ is the estimate, and we are able to come up with a monotone increasing transformation m such that $m(\theta)$ is normally distributed about $m(\hat{\theta})$. We could use this normal distribution to obtain an unbiased confidence interval, and then apply a back-transformation to obtain an almost-unbiased confidence interval.[3]

Even with these modifications, we do not recommend the use of the nonparametric bootstrap with samples of fewer than 100 observations. Simulation studies suggest that with small sample sizes, the coverage is far from exact and the endpoints of the intervals vary widely from one set of bootstrap samples to the next. For example, Tu and Zhang [1992] report that with samples of size 50 taken from a normal distribution, the actual coverage of an interval estimate rated at 90% using the BC_a bootstrap is 88%. When the samples are taken from a mixture of two normal distributions (a not uncommon situation with real-life data sets) the actual coverage is 86%. With samples of only 20 in number, the actual coverage is 80%.

More serious when trying to apply the bootstrap is that the endpoints of the resulting interval estimates may vary widely from one set of bootstrap samples to the next. For example, when Tu and Zhang drew samples of size 50 from a mixture of normal distributions, the average of the left limit of 1000 bootstrap samples taken from each of 1000 simulated data sets was 0.72 with a standard deviation of 0.16, and the average and standard deviation of the right limit were 1.37 and 0.30, respectively.

Parametric Bootstrap

Even when we know the form of the population distribution, the use of the *parametric bootstrap* to obtain interval estimates may prove advantageous either because the parametric bootstrap provides more accurate answers than textbook formulas or because no textbook formulas exist.

Suppose we know that the observations come from a normal distribution and want an interval estimate for the standard deviation. We would draw repeated bootstrap samples from a normal distribution, the mean of which is the sample mean and the variance of which is the sample variance.

---

[3] Stata™ provides for bias-corrected intervals via its bstrap command. R- and S-Plus both include BC_a functions. A SAS macro is available at http://www.asu.edu/it/fyi/research/helpdocs/statistics/SAS/tips/jackboot.html.

As a practical matter, we would draw an element from an $N(0,1)$ population, multiply by the sample standard deviation, and then add the sample mean to obtain an element of our bootstrap sample. By computing the standard deviation of each bootstrap sample, an interval estimate for the standard deviation of the population may be derived.

IMPROVED RESULTS

In many instances, we can obtain narrower interval estimates that have a greater probability of including the true value of the parameter by focusing on sufficient statistics, pivotal statistics, and admissible statistics.

A statistic T is *sufficient* for a parameter if the conditional distribution of the observations given this statistic T is independent of the parameter. If the observations in a sample are exchangeable, then the order statistics of the sample are sufficient; that is, if we know the order statistics $x_{(1)} \le x_{(2)} \le \ldots \le x_{(n)}$, then we know as much about the unknown population distribution as we would if we had the original sample in hand. If the observations are on successive independent binomial trials that end in either success or failure, then the number of successes is sufficient to estimate the probability of success. The minimal sufficient statistic that reduces the observations to the fewest number of discrete values is always preferred.

A *pivotal* quantity is any function of the observations and the unknown parameter that has a probability distribution that does not depend on the parameter. The classic example is Student's t, whose distribution does not depend on the population mean or variance when the observations come from a normal distribution.

A decision procedure d based on a statistic T is *admissible* with respect to a given loss function L, provided that there does not exist a second procedure d^* whose use would result in smaller losses whatever the unknown population distribution.

The importance of admissible procedures is illustrated in an expected way by Stein's paradox. The sample mean, which plays an invaluable role as an estimator of the population mean of a normal distribution for a single set of observations, proves to be inadmissible as an estimator when we have three or more independent sets of observations to work with. Specifically, if $\{X_{ij}\}$ are independent observations taken from four or more distinct normal distributions with means θ_i and variance 1, and losses are proportional to the square of the estimation error, then the estimators

$$\hat{\theta}_i = \overline{X} + (1 - [k-3]/S^2)(\overline{X}_{i.} - \overline{X}_{..}), \qquad \text{where } S^2 = \sum_{i=1}^{k} (\overline{X}_{i.} - \overline{X}_{..})^2,$$

have smaller expected losses than the individual sample means, regardless of the actual values of the population means (see Efron and Morris [1977]).

SUMMARY

Desirable estimators are impartial, consistent, efficient, and robust, and they have minimum loss. Interval estimates are to be preferred to point estimates; they are less open to challenge for they convey information about the estimate's precision.

TO LEARN MORE

Selecting more informative endpoints is the focus of Berger [2002] and Bland and Altman [1995].

Lehmann and Casella [1998] provide a detailed theory of point estimation.

Robust estimators are considered in Huber [1981], Maritz [1996], and Bickel et al. [1993]. Additional examples of both parametric and nonparametric bootstrap estimation procedures may be found in Efron and Tibshirani [1993]. Shao and Tu [1995, Section 4.4] provide a more extensive review of bootstrap estimation methods along with a summary of empirical comparisons.

Carroll and Ruppert [2000] show how to account for differences in variances between populations; this is a necessary step if one wants to take advantage of Stein–James–Efron–Morris estimators.

Bayes estimators are considered in Chapter 6.

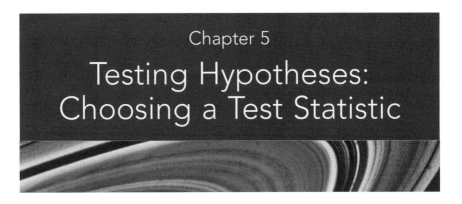

Chapter 5
Testing Hypotheses: Choosing a Test Statistic

"Forget 'large-sample' methods. In the real world of experiments samples are so nearly always 'small' that it is not worth making any distinction, and small-sample methods are no harder to apply." George Dyke [1997].

EVERY STATISTICAL PROCEDURE RELIES ON CERTAIN ASSUMPTIONS FOR correctness. Errors in testing hypotheses come about either because the assumptions underlying the chosen test are not satisfied or because the chosen test is less powerful than other competing procedures. We shall study each of these lapses in turn.

First, virtually all statistical procedures rely on the assumption that the observations are independent.

Second, virtually all statistical procedures require at least one of the following successively weaker assumptions be satisfied under the null hypothesis:

1. The observations are identically distributed.
2. The observations are exchangeable; that is, their joint distribution is the same for any relabeling.
3. The observations are drawn from populations in which a specific parameter is the same across the populations.

The first assumption is the strongest assumption. If it is true, the following two assumptions are also true. The first assumption must be true for a parametric test to provide an exact significance level. If the second assumption is true, the third assumption is also true. The second assumption must be true for a permutation test to provide an exact significance level.

The third assumption is the weakest assumption. It must be true for a bootstrap test to provide an exact significance level asymptotcally.

Common Errors in Statistics (and How to Avoid Them), by Phillip I. Good and James W. Hardin. ISBN 0-471-46068-0 Copyright © 2003 John Wiley & Sons, Inc.

TABLE 5.1 Types of Statistical Tests of Hypotheses

Test Type	Definition	Example
Exact	Stated significance level is exact, not approximate.	t test when observations are i.i.d. normal; permutation test when observations are exchangeable.
Parametric	Obtains cutoff points from specific parametric distribution.	t test
Nonparametric bootstrap	Obtains cutoff points from percentiles of bootstrap distribution of parameter.	
Parametric bootstrap	Obtains cutoff points from percentiles of parameterized bootstrap distribution of parameter.	
Permutation	Obtains cutoff points from distribution of test statistic obtained by rearranging labels.	Tests may be based upon the original observations, on ranks, on normal or Savage scores, or on U statistics.

An immediate consequence of the first two assumptions is that if observations come from a multiparameter distribution, then all parameters, not just the one under test, must be the same for all observations under the null hypothesis. For example, a t test comparing the means of two populations requires that the variation of the two populations be the same.

For nonparametric and parametric bootstrap tests, under the null hypothesis, the observations must all come from a distribution of a specific form.

Let us now explore the implications of these assumptions in a variety of practical testing situations including comparing the means of two populations, comparing the variances of two populations, comparing the means of three or more populations, and testing for significance in two-factor and higher-order experimental designs.

In each instance, before we choose[1] a statistic, we check which assumptions are satisfied, which procedures are most robust to violation of these assumptions, and which are most powerful for a given significance level and sample size. To find the most powerful test, we determine which procedure requires the smallest sample size for given levels of Type I and Type II error.

---

[1] Whether Republican or Democrat, Liberal or Conservative, male or female, we have the right to choose and need not be limited by what textbook, half-remembered teacher pronouncements, or software dictate.

VERIFY THE DATA

The first step in any analysis is to verify that the data have been entered correctly. As noted in Chapter 3, GIGO. A short time ago, a junior biostatistician came into my office asking for help with covariate adjustments for race. "The data for race doesn't make sense," she said. Indeed the proportions of the various races did seem incorrect. No "adjustment" could be made. Nor was there any reason to believe that race was the only variable affected. The first and only solution was to do a thorough examination of the database and, where necessary, trace the data back to its origins until all the bad data had been replaced with good.

The SAS programmer's best analysis tool is PROC MEANS. By merely examining the maximum and minimum values of all variables, it often is possible to detect data that were entered in error. Some years ago, I found that the minimum value of one essential variable was zero. I brought this to the attention of a domain expert who told me that a zero was impossible. As it turns out, the data were full of zeros, the explanation being that the executive in charge had been faking results. Of the 150 subjects in the database, only 50 were real.

Before you begin any analysis, verify that the data have been entered correctly.

COMPARING MEANS OF TWO POPULATIONS

The most common test for comparing the means of two populations is based upon Student's t. For Student's t test to provide significance levels that are exact rather than approximate, all the observations must be independent and, under the null hypothesis, all the observations must come from identical normal distributions.

Even if the distribution is not normal, the significance level of the t test is almost exact for sample sizes greater than 12; for most of the distributions one encounters in practice,[2] the significance level of the t test is usually within a percent or so of the correct value for sample sizes between 6 and 12.

There are more powerful tests than the t test for testing against non-normal alternatives. For example, a permutation test replacing the original observations with their normal scores is more powerful than the t test (Lehmann and D'Abrera, 1988).

Permutation tests are derived by looking at the distribution of values the test statistic would take for each of the possible assignments of treatments to subjects. For example, if in an experiment two treatments were

---

[2] Here and throughout this text, we deliberately ignore the many exceptional cases (to the delight of the true mathematician) that one is unlikely to encounter in the real world.

assigned at random to six subjects so that three subjects got one treatment and three the other, there would have been a total of 20 possible assignments of treatments to subjects.[3] To determine a p value, we compute for the data in hand each of the 20 possible values the test statistic might have taken. We then compare the actual value of the test statistic with these 20 values. If our test statistic corresponds to the most extreme value, we say that $p = 1/20 = 0.05$ (or $1/10 = 0.10$ if this is a two-tailed permutation test).

Against specific normal alternatives, this two-sample permutation test provides a most powerful unbiased test of the distribution-free hypothesis that the centers of the two distributions are the same (Lehmann, 1986, p. 239). For large samples, its power against normal alternatives is almost the same as Student's t test (Albers, Bickel, and van Zwet, 1976). Against other distributions, by appropriate choice of the test statistic, its power can be superior (Lambert, 1985; and Maritz, 1996).

Testing Equivalence

When the logic of a situation calls for demonstration of similarity rather than differences among responses to various treatments, then equivalence tests are often more relevant than tests with traditional no-effect null hypotheses (Anderson and Hauck, 1986; Dixon, 1998; pp. 257–301).

Two distributions F and G such that $G[x] = F[x - \delta]$ are said to be equivalent provided that $|\delta| < \Delta$, where Δ is the smallest difference of clinical significance. To test for equivalence, we obtain a confidence interval for δ, rejecting equivalence *only if* this interval contains valuse in excess of Δ. The width of a confidence interval decreases as the sample size increases; thus a very large sample may be required to demonstrate equivalence just as a very large sample may be required to demonstrate a clinically significant effect.

Unequal Variances

If the variances of the two populations are not the same, neither the t test nor the permutation test will yield exact significance levels despite pronouncements to the contrary of numerous experts regarding the permutation tests.

More important than comparing the means of populations can be determining why the variances are different.

There are numerous possible solutions for the Behrens–Fisher problem of unequal variances in the treatment groups. These include the following:

---

[3] Interested readers may want to verify this for themselves by writing out all the possible addignments of six items into two groups of three, 1 2 3 / 4 5 6, 1 2 4 / 3 5 6, and so forth.

- Wilcoxon test; the use of the ranks in the combined sample reduces the impact (though not the entire effect) of the difference in variability between the two samples.
- Generalized Wilcoxon test (see O'Brien [1988]).
- Procedure described in Manly and Francis [1999].
- Procedure described in Chapter 7 of Weerahandi [1995].
- Procedure described in Chapter 10 of Pesarin [2001].
- Bootstrap. See the section on dependent observations in what follows.
- Permutation test. Phillip Good conducted simulations for sample sizes between 6 and 12 drawn from normally distributed populations. The populations in these simulations had variances that differed by up to a factor of five, and nominal p values of 5% were accurate to within 1.5%.

Hilton [1996] compared the power of the Wilcoxon test, O'Brien's test, and the Smirnov test in the presence of both location shift and scale (variance) alternatives. As the relative influence of the difference in variances grows, the O'Brien test is most powerful. The Wilcoxon test loses power in the face of different variances. If the variance ratio is 4 : 1, the Wilcoxon test is not trustworthy.

One point is unequivocal. William Anderson writes, "The first issue is to understand *why* the variances are so different, and what does this mean to the patient. It may well be the case that a new treatment is not appropriate because of higher variance, even if the difference in means is favorable. This issue is important whether or not the difference was anticipated. Even if the regulatory agency does not raise the issue, I want to do so internally."

David Salsburg agrees. "If patients have been assigned at random to the various treatment groups, the existence of a significant difference in any parameter of the distribution suggests that there is a difference in treatment effect. The problem is not how to compare the means but how to determine what aspect of this difference is relevant to the purpose of the study.

"Since the variances are significantly different, I can think of two situations where this might occur:

1. In many measurements there are minimum and maximum values that are possible, e.g. the Hamilton Depression Scale, or the number of painful joints in arthritis. If one of the treatments is very effective, it will tend to push values into one of the extremes. This will produce a change in distribution from a relatively symmetric one to a skewed one, with a corresponding change in variance.

2. The experimental subjects may represent a mixture of populations. The difference in variance may occur because the

effective treatment is effective for only a subset of the population. A locally most powerful test is given in Conover and Salsburg [1988]."

Dependent Observations

The preceding statistical methods are not applicable if the observations are interdependent. There are five cases in which, with some effort, analysis may still be possible: repeated measures, clusters, known or equal pairwise dependence, a moving average or autoregressive process,[4] and group randomized trials.

Repeated Measures. Repeated measures on a single subject can be dealt with in a variety of ways including treating them as a single multivariate observation. Good [2001, Section 5.6] and Pesarin [2001, Chapter 11] review a variety of permutation tests for use when there are repeated measures.

Another alternative is to use one of the standard modeling approaches such as random- or mixed-effects models or generalized estimating equations (GEEs). See Chapter 10 for a full discussion.

Clusters. Occasionally, data will have been gathered in clusters from families and other groups who share common values, work, or leisure habits. If stratification is not appropriate, treat each cluster as if it were a single observation, replacing individual values with a summary statistic such as an arithmetic average (Mosteller and Tukey, 1977).

Cluster-by-cluster means are unlikely to be identically distributed, having variances, for example, that will depend on the number of individuals that make up the cluster. A permutation test based on these means would not be exact.

If there are a sufficiently large number of such clusters in each treatment group, the *bootstrap* defined in Chapter 3 is the appropriate method of analysis.

With the bootstrap, the sample acts as a surrogate for the population. Each time we draw a pair of bootstrap samples from the original sample, we compute the difference in means. After drawing a succession of such samples, we'll have some idea of what the distribution of the difference in means would be were we to take repeated pairs of samples from the population itself.

As a general rule, resampling should reflect the null hypothesis, according to Young [1986] and Hall and Wilson [1991]. Thus, in contrast to the bootstrap procedure used in estimation (see Chapter 3), each pair of bootstrap samples should be drawn from the *combined sample* taken from

---

[4] For a discussion of these latter, see Brockwell and Davis [1987].

the two treatment groups. Under the null hypothesis, this will not affect the results; under an alternative hypothesis, the two bootstrap sample means will be closer together than they would if drawn separately from the two populations. The difference in means between the two samples that were drawn originally should stand out as an extreme value.

Hall and Wilson [1991] also recommend that the bootstrap be applied only to statistics that, for very large samples, will have distributions that do not depend on any unknowns.[5] In the present example, Hall and Wilson [1991] recommend the use of the t statistic, rather than the simple difference of means, as leading to a test that is both closer to exact and more powerful.

Suppose we draw several hundred such bootstrap samples with replacement from the combined sample and compute the t statistic each time. We would then compare the original value of the test statistic, Student's t in this example, with the resulting bootstrap distribution to determine what decision to make.

Pairwise Dependence. If the covariances are the same for each pair of observations, then the permutation test described previously is an exact test if the observations are normally distributed (Lehmann, 1986) and is almost exact otherwise.

Even if the covariances are not equal, if the covariance matrix is non-singular, we may use the inverse of this covariance matrix to transform the original (dependent) variables to independent (and hence exchangeable) variables. After this transformation, the assumptions are satisfied so that a permutation test can be applied. This result holds even if the variables are collinear. Let R denote the rank of the covariance matrix in the singular case. Then there exists a projection onto an R-dimensional subspace where R normal random variables are independent. So if we have an N dimensional ($N > R$) correlated and singular multivariate normal distribution, there exists a set of R linear combinations of the original N variables so that the R linear combinations are each univariate normal and independent.

The preceding is only of theoretical interest unless we have some independent source from which to obtain an estimate of the covariance matrix. If we use the data at hand to estimate the covariances, the estimates will be interdependent and so will the transformed observations.

Moving Average or Autoregressive Process. These cases are best treated by the same methods and are subject to the caveats as described in Part 3 of this text.

---

[5] Such statistics are termed asymptotically pivotal.

Group Randomized Trials.[6] Group randomized trials (GRTs) in public health research typically use a small number of randomized groups with a relatively large number of participants per group. Typically, some naturally occurring groups are targeted: work sites, schools, clinics, neighborhoods, even entire towns or states. A group can be assigned to either the intervention or control arm but not both; thus, the group is nested within the treatment. This contrasts with the approach used in multicenter clinical trials, in which individuals within groups (treatment centers) may be assigned to any treatment.

GRTs are characterized by a positive correlation of outcomes within a group, along with a small number of groups. "There is positive intraclass correlation (ICC) between the individuals' target-behavior outcomes within the same group. This can be due in part to the differences in characteristics between groups, to the interaction between individuals within the same group, or (in the presence of interventions) to commonalities of the intervention experienced by an entire group. Although the size of the ICC in GRTs is usually very small (e.g., in the Working Well Trial, between 0.01 and 0.03 for the four outcome variables at baseline), its impact on the design and analysis of GRTs is substantial."

"The sampling variance for the average responses in a group is $(\sigma^2/n)*[1 + (n - 1)\sigma)]$, and that for the treatment average with k groups and n individuals per group is $(\sigma^2/n)*[1 + (n - 1)\sigma]$, not the traditional σ^2/n and $\sigma^2/(nk)$, respectively, for uncorrelated data."

"The factor $1 + (n - 1)\sigma$ is called the variance inflation factor (VIF), or design effect. Although σ in GRTs is usually quite small, the VIFs could still be quite large because VIF is a function of the product of the correlation an group size n."

"For example, in the Working Well Trial, with $\sigma = 0.03$ for daily number of fruit and vegetable servings, and an average of 250 workers per work site, VIF = 8.5. In the presence of this deceivingly small ICC, an 8.5-fold increase in the number of participants is required in order to maintain the same statistical power as if there were no positive correlation. Ignoring the VIF in the analysis would lead to incorrect results: variance estimates for group averages that are too small."

To be appropriate, an analysis method of GRTs need to acknowledge both the ICC and the relatively small number of groups. Three primary approaches are used (Table 5.2):

1. *Generalized Linear Mixed Models (GLMM).* **This approach, implemented in SAS Macro GLIMMIX and SAS PROC MIXED, relies on an assumption of normality.**

---

[6] This section has been abstracted (with permission from *Annual Reviews*) from Feng et al. [2001], from whom all quotes in this section are taken.

2. *Generalized Estimating Equations (GEE).* Again, this approach assumes asymptotic normality for conducting inference, a good approximation only when the number of groups is large.

3. *Randomization-Based Inference.* Unequal-sized groups will result in unequal variances of treatment means resulting in misleading p values. To be fair, "Gail et al. [1996] demonstrate that in GRTs, the permutation test remains valid (exact or near exact in nominal levels) under almost all practical situations, including unbalanced group sizes, as long as the number of groups are equal between treatment arms or equal within each block if blocking is used."

The drawbacks of all three methods, including randomization-based inference if corrections are made for covariates, are the same as those for other methods of regression as detailed in Chapters 8 and 9.

TABLE 5.2 Comparison of Different Analysis Methods for Inference on Treatment Effect $\hat{\beta}$[a]

Method $10^2\,\hat{\beta}(10^2\,\mathrm{SE})$	p Value	$\hat{p}$	
Fruit /vegetable			
GLIM (independent)	−6.9 (2.0)	0.0006	
GEE (exchangeable)	−6.8 (2.4)	0.0052	0.0048
GLMM (random intercept)	−6.7 (2.6)	0.023	0.0077
df D 12[b]			
Permutation	−6.1 (3.4)	0.095	
t test (group level)	−6.1 (3.4)	0.098	
Permutation (residual)	−6.3 (2.9)	0.052	
Smoking			
GLIM (independent)	−7.8 (12)	0.53	
GEE (exchangeable)	−6.2 (20)	0.76	0.0185
GLMM (random intercept)	−13 (21)	0.55	0.020
df D 12[b]			
Permutation	−12 (27)	0.66	
t-test (group-level)	−12 (27)	0.66	
Permutation (residual)	−13 (20)	0.53	

[a] Using Seattle 5-a-day data with 26 work sites ($K = 13$) and an average of 87 (n_i ranges from 47 to 105) participants per work site. The dependent variables are ln (daily servings of fruit and vegetable C_1) and smoking status. The study design is matched pair, with two cross-sectional surveys at baseline and 2-year follow-up. Pairs identification, work sites nested within treatment, intervention indicator, and baseline work-site mean fruit-and-vegetable intake are included in the model. Pairs and work sites are random effects in GLMM (generalized linear mixed models). We used SAS PROC GENMOD for GLIM (linear regression and generalized linear models) and GEE (generalized estimating equations) (logistic model for smoking data) and SAS PROCMIXED (for fruit/vegetable data) or GLMMIX (logistic regression for smoking data) for GLMM; permutation tests (logit for smoking data) were programmed in SAS.
[b] Degrees of freedom (df) = 2245 in SAS output if work site is not defined as being nested within treatment.
Source: Reprinted with permission from the *Annual Review of Public Health* Volume 22, © 2001 by Annual Reviews. Feng et al. [2002].

Nonsystematic Dependence. If the observations are interdependent and fall into none of the preceding categories, then the experiment is fatally flawed. Your efforts would be best expended on the design of a cleaner experiment. Or, as J. W. Tukey remarked on more than one occasion, "If a thing is not worth doing, it is not worth doing well."

COMPARING VARIANCES

Testing for the equality of the variances of two populations is a classic problem with many not-quite-exact, not-quite-robust, not-quite-powerful-enough solutions. Sukhatme [1958] lists four alternative approaches and adds a fifth of his own; Miller [1968] lists 10 alternatives and compares four of these with a new test of his own; Conover, Johnson, and Johnson [1981] list and compare 56 tests; and Balakrishnan and Ma [1990] list and compare nine tests with one of their own.

None of these tests proves satisfactory in all circumstances, because each requires that two or more of the following four conditions be satisfied:

1. The observations are normally distributed.
2. The location parameters of the two distributions are the same or differ by a known quantity.
3. The two samples are equal in size.
4. The samples are large enough that asymptotic approximations to the distribution of the test statistic are valid.

As an example, the first published solution to this classic testing problem is the z test proposed by Welch [1937] based on the ratio of the two sample variances. If the observations are normally distributed, this ratio has the F distribution, and the test whose critical values are determined by the F distribution is uniformly most powerful among all unbiased tests (Lehmann, 1986, Section 5.3). But with even small deviations from normality, significance levels based on the F distribution are grossly in error (Lehmann, 1986, Section 5.4).

Box and Anderson [1955] propose a correction to the F distribution for "almost" normal data, based on an asymptotic approximation to the permutation distribution of the F ratio. Not surprisingly, their approximation is close to correct only for normally distributed data or for very large samples. The Box–Anderson statistic results in an error rate of 21%, twice the desired value of 10%, when two samples of size 15 are drawn from a gamma distribution with four degrees of freedom.

A more recent permutation test (Bailor, 1989) based on complete enumeration of the permutation distribution of the sample F ratio is exact only when the location parameters of the two distributions are known or are known to be equal.

The test proposed by Miller [1968] yields conservative Type I errors, less than or equal to the declared error, unless the sample sizes are unequal. A 10% test with samples of size 12 and 8 taken from normal populations yielded Type I errors 14% of the time.

Fligner and Killeen [1976] propose a permutation test based on the sum of the absolute deviations from the combined sample mean. Their test may be appropriate when the medians of the two populations are equal, but can be virtually worthless otherwise, accepting the null hypothesis up to 100% of the time. In the first edition, Good [2001] proposed a test based on permutations of the absolute deviations from the individual sample medians; this test, alas, is only asymptotically exact and even then only for approximately equal sample sizes, as shown by Baker [1995].

To compute the primitive bootstrap introduced by Efron [1979], we would take successive pairs of samples—one of n observations from the sampling distribution F_n which assigns mass $1/n$ to the values $\{X_i: i = 1, \ldots, n\}$, and one of m observations from the sampling distribution G_m which assigns mass $1/m$ to the values $\{X_j: j = n + 1, \ldots, n + m\}$, and compute the ratio of the sample variances

$$R = \frac{s_n^2/(n-1)}{s_m^2/(m-1)}.$$

We would use the resultant bootstrap distribution to test the hypothesis that the variance of F equals the variance of G against the alternative that the variance of G is larger. Under this test, we reject the null hypothesis if the $100(1 - \alpha)$ percentile is less than 1.

This primitive bootstrap and the associated confidence intervals are close o exact only for very large samples with hundreds of observations. More often the true coverage probability is larger than the desired value.

Two corrections yield vastly improved results. First, for unequal-sized samples, Efron [1982] suggests that more accurate confidence intervals can be obtained using the test statistic

$$R' = \frac{s_n^2/n}{s_m^2/m}$$

Second, applying the bias and acceleration corrections described in Chapter 3 to the bootstrap distribution of R' yields almost exact intervals.

Lest we keep you in suspense, a distribution-free exact and more powerful test for comparing variances can be derived based on the permutation distribution of Aly's statistice.

This statistic proposed by Aly [1990] is

$$\delta = \sum_{i=1}^{m-1} i(m-i)(X_{(i+1)} - X_{(i)})$$

where $X_{(1)} \le X_{(2)} \le \ldots \le X_{(m)}$ are the order statistics of the first sample.

Suppose we have two sets of measurements, 121, 123, 126, 128.5, 129 and in a second sample, 153, 154, 155, 156, 158. We replace these with the deviations $z_{1i} = X_{(i+1)} - X_{(i)}$ or 2, 3, 2.5, .5 for the first sample and $z_{2i} = 1, 1, 1, 2$ for the second.

The original value of the test statistic is $8 + 18 + 15 + 2 = 43$. Under the hypothesis of equal dispersions in the two populations, we can exchange labels between z_{1i} and z_{2i} for any or all of the values of i. One possible rearrangement of the labels on the deviations puts {2, 1, 1, 2} in the first sample, which yields a value of $8 + 6 + 6 + 8 = 28$.

There are $2^4 = 16$ rearrangements of the labels in all, of which only one {2, 3, 2.5, 2} yields a larger value of Aly's statistic than the original observations. A one-sided test would have two out of 16 rearrangements as or more extreme than the original, and a two-sided test would have four. In either case, we would accept the null hypothesis, though the wiser course would be to defer judgment until we have taken more observations.

If our second sample is larger than the first, we have to resample in two stages. First, we select a subset of m values at random without replacement from the n observations in the second, larger sample and compute the order statistics and their differences. Last, we examine all possible values of Aly's measure of dispersion for permutations of the combined sample as we did when the two samples were equal in size and compare Aly's measure for the original observations with this distribution. We repeat this procedure several times to check for consistency.

COMPARING THE MEANS OF K SAMPLES

The traditional one-way analysis of variance based on the F ratio

$$\frac{\sum_{i=1}^{I} n_i (X_{i.} - X_{..})^2 / (I-1)}{\sum_{i=1}^{I} \sum_{j=1}^{n_i} (X_{ij} - X_{i.})^2 / (N-I)}$$

has at least three major limitations:

MATCH SIGNIFICANCE LEVELS BEFORE PERFORMING POWER COMPARISONS

When we studied the small-sample properties of parametric tests based on asymptotic approximations that had performed well in previously published power comparisons, we uncovered another major error in statistics: the failure to match significance levels before performing power comparisons. Asymptotic approximations to cutoff value were used rather than exact values or near estimates.

When a statistical test takes the form of an interval, that is, if we reject when $S < c$ and accept otherwise, then power is a nondecreasing function of significance level; a test based on an interval may have greater power at the 10% significance level than a second different test evaluated at the 5% significance level, even though the second test is uniformly more powerful than the first. To see this, let H denote the primary hypothesis and let K denote an alternative hypothesis:

If $\Pr\{S < c|H\} = \alpha < \alpha' = \Pr\{S < c'|H\}$, then $c < c'$, and $\beta = \Pr\{S < c|K\} \leq \Pr\{S < c'|K\} = \beta'$.

Consider a second statistical test depending on S via the monotone increasing function h, where we reject if $h[S] < d$ and accept otherwise. If the cutoff values $d < d'$ correspond to the same significance levels $\alpha < \alpha'$, then $\beta < \Pr\{h[S] < d|K\} < \beta'$. Even though the second test is more powerful than the first at level α, this will not be apparent if we substitute an approximate cutoff point c' for an exact one c when comparing the two tests.

To ensure matched significance levels in your own power comparisons, proceed in two stages: First, use simulations to derive exact cutoff values. Then, use these derived cutoff values in determining power. Using this approach, we were able to show that an exact permutation test based on Aly's statistic was more powerful for comparing variances than any of the numerous published inexact parametric tests.

1. Its significance level is heavily dependent on the assumption of normality.

2. The F ratio is optimal for losses that are proportional to the square of the error and is suboptimal otherwise.

3. The F ratio is an omnibus statistic offering all-round power against many alternatives but no particular advantage against any specific one of them.

"Normality is a myth; there never has, and never will be a normal distribution."

Geary[1947, p. 241]

A permutation test is preferred for the k-sample analysis. These tests are distribution-free (though the variances must be the same for all treatments). And you can choose the test statistic that is optimal for a given alternative and loss function and not be limited by the availability of tables.

We take as our model $X_{ij} = \mu + \alpha_i + \varepsilon_{ij}$, where we select μ so that the treatment effects α_i sum to zero; $i = 1, \ldots, I$ denotes the treatment, and $j = 1, \ldots, n_i$. We assume that the error terms $\{\varepsilon_{ij}\}$ are independent and identically distributed.

We consider two loss functions: one in which the losses associated with overlooking a real treatment effect, a Type II error, are proportional to the sum of the squares of the treatment effects α_i^2 (LS), the other in which the losses are proportional to the sum of the absolute values of the treatment effects, $|\alpha_i|$ (LAD).

Our hypothesis, a null hypothesis, is that the differential treatment effects, the $\{\alpha_i\}$, are all zero. We will also consider two alternative hypotheses: K_U that at least one of the differential treatment effects α_i is not zero, and K_O that K_U is true and there is an ordered response such that $\alpha_1 \leq \alpha_2 \leq \ldots \leq \alpha_I$.

For testing against K_U with the LS loss function, Good [2002, p. 126] recommends the use of the statistic $F_2 = \Sigma_i(\Sigma_j X_{ij})^2$ which is equivalent to the F ratio once terms that are invariant under permutations are eliminated.

For testing against K_U with the LAD loss function, Good [2002, p. 126] recommends the use of the statistic $F_1 = \Sigma_i|\Sigma_j X_{ij}|$.

For testing against K_O, Good [2001, p. 46] recommends the use of the Pitman correlation statistic $\Sigma_i f[i]\Sigma_j X_{ij}$, where $f[i]$ is a monotone increasing function of i that depends upon the alternative. For example, for testing for a dose response in animals where i denotes the dose, one might use $f[i] = \log[i + 1]$.

A permutation test based on the original observations is appropriate only if one can assume that under the null hypothesis the observations are identically distributed in each of the populations from which the samples are drawn. If we cannot make this assumption, we will need to transform the observations, throwing away some of the information about them so that the distributions of the transformed observations are identical.

For example, for testing against K_O, Lehmann [1999, p. 372] recommends the use of the Jonckheere–Terpstra statistic, the number of pairs in which an observation from one group is less than an observation from a higher-dose group. The penalty we pay for using this statistic and ignoring the actual values of the observations is a marked reduction in power for small samples and is a less pronounced loss for larger ones.

If there are just two samples, the test based on the Jonckheere–Terpstra statistic is identical to the Mann–Whitney test. For very large samples, with identically distributed observations in both samples, 100 observations would be needed with this test to obtain the same power as a permutation test based on the original values of 95 observations. This is not a price one would want to pay in human or animal experiments.

HIGHER-ORDER EXPERIMENTAL DESIGNS

Similar caveats hold for the parametric ANOVA approach to the analysis of two-factor experimental design with two additions:

1. **The sample sizes must be the same in each cell; that is, the design must be balanced.**
2. **A test for interaction must precede any test for main effects.**

Imbalance in the design will result in the confounding of main effects with interactions. Consider the following two-factor model for crop yield:

$$X_{ijk} = \mu + \alpha_i + \beta_j + \gamma_{ij} + \varepsilon_{jjk}$$

Now suppose that the observations in a two-factor experimental design are normally distributed as in the following diagram taken from Cornfield and Tukey (1956):

$$\frac{N(0,1)\,|\,N(2,1)}{N(2,1)\,|\,N(0,1)}$$

There are no main effects in this example—both row means and both column means have the same expectations, but there is a clear interaction represented by the two nonzero off-diagonal elements.

If the design is balanced, with equal numbers per cell, the lack of significant main effects and the presence of a significant interaction should and will be confirmed by our analysis. But suppose that the design is not in balance, that for every 10 observations in the first column, we have only one observation in the second. Because of this imbalance, when we use the F ratio or equivalent statistic to test for the main effect, we will uncover a false "row" effect that is actually due to the interaction between rows and columns. The main effect is *confounded* with the interaction.

If a design is unbalanced as in the preceding example, we cannot test for a "pure" main effect or a "pure" interaction. But we may be able to test for the combination of a main effect with an interaction by using the statistic that we would use to test for the main effect alone. This combined effect will not be confounded with the main effects of other unrelated factors.

Whether or not the design is balanced, the presence of an interaction may zero out a cofactor-specific main effect or make such an effect impossible to detect. More important, the presence of a significant interaction may render the concept of a single "main effect" meaningless. For example, suppose we decide to test the effect of fertilizer and sunlight on plant growth. With too little sunlight, a fertilizer would be completely ineffective. Its effects only appear when sufficient sunlight is present. Aspirin and warfarin can both reduce the likelihood of repeated heart attacks when used alone; you don't want to mix them!

Gunter Hartel offers the following example: Using five observations per cell and random normals as indicated in Cornfield and Tukey's diagram, a two-way ANOVA without interaction yields the following results:

Source	df	Sum of Squares	F Ratio	Prob > F
Row	1	0.15590273	0.0594	0.8104
Col	1	0.10862944	0.0414	0.8412
Error	17	44.639303		

Adding the interaction term yields

Source	df	Sum of Squares	F Ratio	Prob > F
Row	1	0.155903	0.1012	0.7545
Col	1	0.108629	0.0705	0.7940
Row*col	1	19.986020	12.9709	0.0024
Error	16	24.653283		

Expanding the first row of the experiment to have 80 observations rather than 10, the main effects only table becomes

Source	df	Sum of Squares	F Ratio	Prob > F
Row	1	0.080246	0.0510	0.8218
Col	1	57.028458	36.2522	<.0001
Error	88	138.43327		

But with the interaction term it is:

Source	df	Sum of Squares	F Ratio	Prob > F
Row	1	0.075881	0.0627	0.8029
Col	1	0.053909	0.0445	0.8333
row*col	1	33.145790	27.3887	<.0001
Error	87	105.28747		

Independent Tests

Normally distributed random variables (as in Figure 7.1) have some remarkable properties:

- The sum (or difference) of two independent normally distributed random variables is a normally distributed random variable.
- The square of a normally distributed random variable has the chi-square distribution (to within a multiplicative constant); the sum of two variables with the chi-square distribution also has a chi-square distribution (with additional degrees of freedom).
- A variable with the chi-square distribution can be decomposed into the sum of several independent chi-square variables.

As a consequence of these properties, the variance of a sum of independent normally distributed random variables can be decomposed into the sum of a series of independent chi-square variables. We use these independent variables in the analysis of variance (ANOVA) to construct a series of independent tests of the model parameters.

Unfortunately, even slight deviations from normality negate these properties; not only are ANOVA p values in error because they are taken from the wrong distribution, but they are in error because the various tests are interdependent.

When constructing a permutation test for multifactor designs, we must also proceed with great caution for fear that the resulting tests will be interdependent.

The residuals in a two-way complete experimental design are not exchangeable even if the design is balanced as they are both correlated and functions of all the data (Lehmann and D'Abrera, 1988). To see this, suppose our model is $X_{ijk} = \mu + \alpha_i + \beta_j + \gamma_{ij} + \varepsilon_{ijk}$, where $\Sigma \alpha_i = \Sigma \beta_j = \Sigma_i \gamma_{ij} = \Sigma_j \gamma_{ij} = 0$.

Eliminating the main effects in the traditional manner, that is, setting $X'_{ijk} = X_{ijk} - \overline{X}_{i..} - \overline{X}_{.j.} + \overline{X}_{...}$, one obtains the test statistic

$$ I = \sum_i \sum_j \left(\sum_k X'_{ijk} \right)^2 $$

first derived by Still and White [1981]. A permutation test based on the statistic I will not be exact because even if the error terms $\{\varepsilon_{ijk}\}$ are exchangeable, the residuals $X'_{ijk} = \varepsilon_{ijk} - \overline{\varepsilon}_{i..} - \overline{\varepsilon}_{.j.} + \overline{\varepsilon}_{...}$ are weakly correlated, with the correlation depending on the subscripts.

Nonetheless, the literature is filled with references to permutation tests for the two-way and higher-order designs that produce misleading values. Included in this category are those permutation tests based on the ranks of the observations that may be found in many statistics software packages.

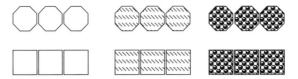

FIGURE 5.1 A 2 × 3 Design with Three Observations per Cell.

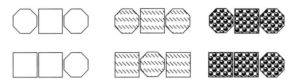

FIGURE 5.2 A 2 × 3 Design with Three Observations per Cell after $\pi \in P_R$.

The recent efforts of Salmaso [2003] and Pesarin [2001] have resulted in a breakthrough that extends to higher-order designs. The key lies in the concept of *weak exchangeability* with respect to a subset of the possible permutations. The simplified discussion of weak exchangeability presented here is abstracted from Good [2003].

Think of the set of observations $\{X_{ijk}\}$ in terms of a rectangular lattice L with K colored, shaped balls at each vertex. All the balls in the same column have the same color initially, a color which is distinct from the color of the balls in any other column. All the balls in the same row have the same shape initially, a shape which is distinct from the shape of the balls in any other row. See Fig. 5.1.

Let P denote the set of rearrangements or permutations that preserve the number of balls at each row and column of the lattice. P is a group.[7]

Let P_R denote the set of exchanges of balls among rows and within columns which (a) preserve the number of balls at each row and column of the lattice and (b) result in the numbers of each shape within each row being the same in each column. P_R is the basis of a subgroup of P. See Fig. 5.2.

Let P_C denote the set of exchanges of balls among columns and within rows which (a) preserve the number of balls at each row and column of the lattice and (b) result in the numbers of each color within each column being the same in each row. P_C is the basis of a subgroup of P. See Fig. 5.3.

Let P_{RC} denote the set of exchanges of balls that preserve the number of balls at each row and column of the lattice and which result in (a) an

---

[7] See Hungerford [1974] or http://members.tripod.com/~dogschool/ for a thorough discussion of algebraic group properties.

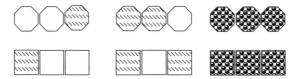

FIGURE 5.3 A 2 × 3 Design with Three Observations per Cell $\pi \in P_c$.

exchange of balls between both rows and columns (or no exchange at all), (b) the numbers of each color within each column being the same in each row, and (c) the numbers of each shape within each row being the same in each column. P_{RC} is the basis of a subgroup of P.

The only element these three subgroups P_{RC}, P_R, and P_C have in common is the rearrangement that leaves the observations with the same row and column labels they had to begin with. As a result, tests based on these three different subsets of permutations are independent of one another.

For testing H_3: $\gamma_{ij} = 0$ for all i and j, determine the distribution of the values of $S = \Sigma_{1 \le i \le I_1} \Sigma_{1 \le j \le I_2} (X_{ij} + X_{i'j'} - X_{i'j} - X_{ij'})$ with respect to the rearrangements in P_{RC}. If the value of S for the observations as they were originally labeled is *not* an extreme value of this permutation distribution, then we can accept the hypothesis H_3 of no interactions and proceed to test for main effects.

For testing H_1: $\alpha_i = 0$ for all i, choose one of the following test statistics as we did in the section on one-way analysis, $F_{12} = \Sigma_i (\Sigma_j \Sigma_k x_{ijk})^2$, $F_{11} = \Sigma_i |\Sigma_j \Sigma_k x_{ijk}|$, or $R_1 = \Sigma_i g[i] \Sigma_j \Sigma_k x_{ijk}$, where $g[i]$ is a monotone function of i, and determine the distribution of its values with respect to the rearrangements in P_R.

For testing H_2: $\beta_j = 0$ for all j, choose one of the following test statistics as we did in the section on one-way analysis, $F_{22} = \Sigma_j (\Sigma_i \Sigma_k x_{ijk})^2$, $F_{21} = \Sigma_j |\Sigma_i \Sigma_k x_{ijk}|$, or $R_2 = \Sigma_j g[j] \Sigma_i \Sigma_k x_{ijk}$, where $g[j]$ is a monotone function of j, and determine the distribution of its values with respect to the rearrangements in P_C.

Tests for the parameters of three-way and higher-order experimental designs can be obtained via the same approach; use a multidimensional lattice and such additional multivalued properties of the balls as charm and spin. Proofs may be seen at http://users.oco.net/drphilgood/resamp.htm.

Unbalanced Designs

Unbalanced designs with unequal numbers per cell may result from unanticipated losses during the conduct of an experiment or survey (or from an extremely poor initial design). There are two approaches to their analysis:

Permutation tests can be applied to unbalanced as well as balanced experimental designs, providing only that are sufficient observations in each cell to avoid confounding of the main effects and interactions. Even in this latter case, exact permutation tests are available; see Pesarin [2001, p. 237], observations, recognizing that the results may be somewhat tainted.

Second, we might bootstrap along one of the following lines:

- **If only one or two observations are missing, create a balanced design by discarding observations at random; repeat to obtain a distribution of *p* values (Baker, 1995).**
- **If there are actual holes in the design, so that there are missing combinations, create a test statistic that does not require the missing data. Obtain its distribution by bootstrap means. See Good [2000, pp. 68–70] for an example.**

CONTINGENCY TABLES

A major source of error in the analysis of contingency tables is to associate the Pearson chi-square statistic, a quite useful measure of the difference between observed and expected values, with the chi-square distribution. The latter is the distribution of Z^2, where Z has the normal distribution.

Just as the means of very large samples have almost normal distributions, so the means of very large numbers of squared values tend to almost chi-square distributions. Pearson's chi-square statistic is no exception to the rule. If the probabilities of an observation falling in a particular cell of a contingency table are roughly the same for all rows and columns, then convergence the chi-square distribution can be quite rapid. But for sparse tables, the chi-square distribution can be quite misleading (Delucchi, 1983).

We recommend using an exact permutation procedure, particularly now that software for a variety of testing situations is commercially and freely available.[8] As in Fisher [1935], we determine the proportion of tables with the same marginals that are as extreme as, or more extreme than, our original table.

The problem lies in defining what is meant by "extreme." The errors lie in failing to report how we arrived at our definition.

For example, in obtaining a two-tailed test for independence in a 2×2 contingency table, we can treat each table strictly in accordance with its probability under the multinomial distribution (Fisher's method) or weight each table by the value of the Pearson chi-square statistic for that table. The situation is even more complicated with general $R \times C$ tables where a dozen different statistics compete for our attention.

---

[8] Examples include StatXact® from http://www.cytel.com, RT from www.west-inc.com, NPC Test from http://www.methodologica.it., and R (freeware) from http://www.r-project.org.

The chief errors in practice lie in failing to report all of the following:

- **Whether we used a one-tailed or two-tailed test and why.**
- **Whether the categories are ordered or unordered.**
- **Which statistic was employed and why.**

Chapter 9 contains a discussion of a final, not inconsiderable source of error, the neglect of confounding variables that may be responsible for creating an illusory association or concealing an association that actually exists.

INFERIOR TESTS

Violation of assumptions can affect not only the significance level of a test but the power of the test, as well; see Tukey and McLaughlin [1963] and Box and Tiao [1964]. For example, while the significance level of the t test is robust to departures from normality, the power of the t test is not. Thus, the two-sample permutation test may always be preferable.

If blocking including matched pairs was used in the original design, then the same division into blocks should be employed in the analysis. Confounding factors such as sex, race, and diabetic condition can easily mask the effect we hoped to measure through the comparison of two samples. Similarly, an overall risk factor can be totally misleading (Gigerenzer, 2002). Blocking reduces the differences between subjects so that differences between treatment groups stand out—that is , if the appropriate analysis is used. Thus, paired data should always be analyzed with the paired t test or its permutation equivalent, not with the group t test.

To analyze a block design (for example, where we have sampled separately from whites, blacks, and Hispanics), the permutation test statistic is $S = \Sigma_{b=1}^{B}\Sigma_{j}x_{bj}$, where x_{bj} is the jth observation in the control sample in the bth block, and the rearranging of labels between control and treated samples takes place separately and independently within each of the B blocks (Good, 2001, p. 124).

Blocking can also be used after the fact if you suspect the existence of confounding variables and if you measured the values of these variables as you were gathering data.[9]

Always be sure your choice of statistic is optimal against the alternative hypotheses of interest for the appropriate loss function.

To avoid using an inferior less sensitive and possibly inaccurate statistical procedure, pay heed to another admonition from George Dyke [1997]: "The availability of 'user-friendly' statistical software has caused authors to become increasingly careless about the logic of interpreting their results,

---

[9] This recommendation applies only to a test of efficacy for all groups (blocks) combined. p values for subgroup analyses performed after the fact are still suspect; see Chapter 1.

and to rely uncritically on computer output, often using the 'default option' when something a little different (usually, but not always, a little more complicated) is correct, or at least more appropriate."

MULTIPLE TESTS

When we perform multiple tests in a study, there may not be journal room (nor interest) to report all the results, but we do need to report the total number of statistical tests performed so that readers can draw their own conclusions as to the significance of the results that are reported.

We may also wish to correct the reported significance levels by using one of the standard correction methods for independent tests (e.g., Bonferroni; for resampling methods, see Westfall and Young, 1993).

Several statistical packages—SAS is a particular offender—print out the results of several dependent tests performed on the same set of data—for example, the t test and the Wilcoxon. We are not free to pick and choose. Before we view the printout, we must decide which test we will employ.

Let W_α denote the event that the Wilcoxon test rejects a hypothesis at the α significance level. Let P_α denote the event that a permutation test based on the original observations and applied to the same set of data rejects a hypothesis at the α significance level. Let T_α denote the event that a t test applied to the same set of data rejects a hypothesis at the α significance level.

It is possible that W_α may be true when P_α and T_α are not, and so forth. As $\Pr \{W_\alpha \text{ or } P_\alpha \text{ or } T_\alpha | H\} \leq \Pr \{W_\alpha | H\} = \alpha$, we will have inflated the Type I error by picking and choosing after the fact which test to report. Vice versa, if our intent was to conceal a side effect by reporting that the results were not significant, we will inflate the Type II error and deflate the power β of our test, by an after-the-fact choice as $\beta = \Pr \{\text{not } (W_\alpha \text{ and } P_\alpha \text{ and } T_\alpha) | K\} \leq \Pr\{W_\alpha | K\}$.

To repeat, we are not free to pick and choose among tests; any such conduct is unethical. **Both the comparison and the test statistic must be specified in advance of examining the data.**

BEFORE YOU DRAW CONCLUSIONS

Before you draw conclusions, be sure you have accounted for all missing data, interviewed nonresponders, and determined whether the data were missing at random or were specific to one or more subgroups.

During the Second World War, a group was studying planes returning from bombing Germany. They drew a rough diagram showing where the bullet holes were and recommended those areas be reinforced. A statistician, Abraham Wald [1980],[10] pointed out that essential data were missing from the sample they were studying. What about the planes that didn't return from Germany?

When we think along these lines, we see that the two areas of the plane that had almost no bullet holes (where the wings and where the tail joined the fuselage) are crucial. Bullet holes in a plane are likely to be at random, occurring over the entire plane. Their absence in those two areas in returning bombers was diagnostic. Do the data missing from your experiments and surveys also have a story to tell?

Induction

> *"Behold! human beings living in an underground den, which has a mouth open towards the light and reaching all along the den; here they have been from their childhood, and have their legs and necks chained so that they cannot move, and can only see before them, being prevented by the chains from turning round their heads. Above and behind them a fire is blazing at a distance, and between the fire and the prisoners there is a raised way; and you will see, if you look, a low wall built along the way, like the screen which marionette players have in front of them, over which they show the puppets."*
>
> *"And they see only their own shadows, or the shadows of one another, which the fire throws on the opposite wall of the cave."*
>
> *"To them, I said, the truth would be literally nothing but the shadows of the images."*

The Allegory of the Cave (Plato, The Republic, Book VII).

Never assign probabilities to the true state of nature, but only to the validity of your own predictions.

A *p* value does not tell us the probability that a hypothesis is true, nor does a significance level apply to any specific sample; the latter is a characteristic of our testing in the long run. Likewise, if all assumptions are satisfied, a confidence interval will in the long run contain the true value of the parameter a certain percentage off the time. But we cannot say with certainty in any specific case that the parameter does or does not belong to that interval (Neyman, 1961, 1977).

When we determine a *p* value, we apply a set of algebraic methods and deductive logic to *deduce* the correct value. The deductive process is used

---

[10] This reference may be hard to obtain. Alternatively, see Mangel and Samaniego [1984].

to determine the appropriate size of resistor to use in an electric circuit, to determine the date of the next eclipse of the moon, and to establish the identity of the criminal (perhaps from the fact the dog did not bark on the night of the crime). Find the formula, plug in the values, turn the crank, and out pops the result (or it does for Sherlock Holmes,[11] at least).

When we assert that for a given population that a percentage of samples will have a specific composition, this is a deduction also. But when we make an *inductive* generalization about a population based upon our analysis of a sample, we are on shakier ground. Newton's Law of gravitation provided an exact fit to observed astronomical data for several centuries; consequently, there was general agreement that Newton's generalization from observation was an accurate description of the real world. Later, as improvements in astronomical measuring instruments extended the range of the observable universe, scientists realized that Newton's Law was only a generalization and not a property of the universe at all. Einstein's Theory of Relativity gives a much closer fit to the data, a fit that has not been contradicted by any observations in the century since its formulation. But this still does not mean that relativity provides us with a complete, correct, and comprehensive view of the universe.

In our research efforts, the only statements we can make with God-like certainty are of the form "our conclusions fit the data." The true nature of the real world is unknowable. We can speculate, but never conclude.

The gap between the sample and the population will always require a leap of faith. We understand only in so far as we are capable of understanding [Lonergan, 1992].

SUMMARY

Know your objectives in testing. Know your data's origins. Know the assumptions you feel comfortable with. Never assign probabilities to the true state of nature, but only to the validity of your own predictions. Collecting more and better data may be your best alternative.

TO LEARN MORE

For commentary on the use of wrong or inappropriate statistical methods, see Avram et al. [1985], Badrick and Flatman [1999], Berger et al. [2002], Bland and Altman [1995], Cherry [1998], Dar, Serlin, and Omer [1997], Elwood [1998], Felson, Cupples, and Meenan [1984], Fienberg [1990], Gore, Jones, and Rytter [1977], Lieberson [1985], MacArthur

---

[11] See "Silver Blaze" by A. Conan-Doyle, *Strand Magazine*, December 1892.

and Jackson [1984], McGuigan [1995], McKinney et al. [1989], Miller [1986], Padaki [1989], Welch and Gabbe [1996], Westgard and Hunt [1973], White [1979], and Yoccuz [1991].

Guidelines for reviewers are provided by Altman [1998a], Bacchetti [2002], Finney [1997], Gardner, Machin, and Campbell [1986], George [1985], Goodman, Altman, and George [1998], International Committee of Medical Journal Editors [1997], Light and Pillemer [1984], Mulrow [1987], Murray [1988], Schor and Karten [1966], and Vaisrub [1985].

For additional comments on the effects of the violation of assumptions, see Box and Anderson [1955], Friedman [1937], Gastwirth and Rubin [1971], Glass, Peckham, and Sanders [1972], and Pettitt and Siskind [1981].

For the details of testing for equivalence, see Dixon [1998]. For a review of the appropriate corrections for multiple tests, see Tukey [1991].

For procedures with which to analyze factorial and other multifactor experimental designs, see Chapter 8 of Pesarin [2001].

Most of the problems with parametric tests reported here extend to and are compounded by multivariate analysis. For some solutions, see Chapter 5 of Good [2000] and Chapter 6 of Pesarin [2001].

For a contrary view on the need for adjustments of p values in multiple comparisons, see Rothman [1990a].

Venn [1888] and Reichenbach [1949] are among those who've attempted to construct a mathematical bridge between what we observe and the reality that underlies our observations. To the contrary, extrapolation from the sample to the population is not a matter of applying Holmes-like deductive logic but entails a leap of faith. A careful reading of Locke [1700], Berkeley [1710], Hume [1748], and Lonergan [1992] is an essential prerequisite to the application of statistics.

For more on the contemporary view of induction, see Berger [2002] and Sterne, Smith, and Cox [2001]. The former notes that, "Dramatic illustration of the non-frequentist nature of p-values can be seen from the applet available at http://www.stat.duke.edu/~berger. The applet assumes one faces a series of situations involving normal data with unknown mean θ and known variance, and tests of the form $H: \theta = 0$ versus $K: \theta \neq 0$. The applet simulates a long series of such tests, and records how often H is true for p-values in given ranges."

Chapter 6
Strengths and Limitations of Some Miscellaneous Statistical Procedures

THE GREATEST ERROR ASSOCIATED WITH THE USE OF statistical procedures is to make the assumption that one single statistical methodology can suffice for all applications.

From time to time, a new statistical procedure will be introduced or an old one revived along with the assertion that at last the definitive solution has been found. As is so often the case with religions, at first the new methodology is reviled, even persecuted, until it grows in the number of its adherents, at which time it can begin to attack and persecute the adherents of other, more established dogma in its turn.

During the preparation of this text, an editor of a statistics journal rejected an article of one of the authors on the sole grounds that it made use of permutation methods.

"I'm amazed that anybody is still doing permutation tests . . ." wrote the anonymous reviewer, "There is probably nothing wrong technically with the paper, but I personally would reject it on grounds of irrelevance to current best statistical practice." To which the editor sought fit to add, "The reviewer is interested in estimation of interaction or main effects in the more general semiparametric models currently studied in the literature. It is well known that permutation tests preserve the significance level but that is all they do is answer yes or no."[1]

But one methodology can never be better than another, nor can estimation replace hypothesis testing or vice versa. Every methodology has a proper domain of application and another set of applications for which it

---

[1] A double untruth. First, permutation tests also yield interval estimates; see, for example, Garthwaite [1996]. Second, semiparametric methods are not appropriate for use with small-sample experimental designs, the topic of the submission.

Common Errors in Statistics (and How to Avoid Them), by Phillip I. Good and James W. Hardin.
ISBN 0-471-46068-0   Copyright © 2003 John Wiley & Sons, Inc.

fails. Every methodology has its drawbacks and its advantages, its assumptions, and its sources of error. Let us seek the best from each statistical procedure.

The balance of this chapter is devoted to exposing the frailties of four of the "new" (and revived) techniques: bootstrap, Bayesian methods, meta-analysis, and permutation tests.

BOOTSTRAP

Many of the procedures discussed in this chapter fall victim to the erroneous perception that one can get more out of a sample or series of samples than one actually puts in. One bootstrap expert learned he was being considered for a position because management felt, "your knowledge of the bootstrap will help us to reduce the cost of sampling."

Michael Chernick, author of *Bootstrap Methods: A Practitioner's Guide*, Wiley, 1999, has documented six myths concerning the bootstrap:

1. **Allows you to reduce your sample size requirements by replacing real data with simulated data—Not.**
2. **Allows you to stop thinking about your problem, the statistical design and probability model—Not.**
3. **No assumptions necessary—Not.**
4. **Can be applied to any problem—Not.**
5. **Only works asymptotically—Necessary sample size depends on the context.**
6. **Yields exact significance levels—Never.**

Of course, the bootstrap does have many practical applications, as witnessed by its appearance in six of the chapters in this book.[2]

Limitations

As always, to use the bootstrap or any other statistical methodology effectively, one has to be aware of its limitations. The bootstrap is of value in any situation in which the sample can serve as a surrogate for the population.

If the sample is not representative of the population because the sample is small or biased, not selected at random, or its constituents are not independent of one another, then the bootstrap will fail.

Canty et al. [2000] also list data outliers, inconsistency of the bootstrap method, incorrect resampling model, wrong or inappropriate choice of statistic, nonpivotal test statistics, nonlinearity of the test statistic, and discreteness of the resample statistic as potential sources of error.

---

[2] If you're counting, we meet the bootstrap again in Chapters 10 and 11.

One of the first proposed uses of the bootstrap, illustrated in Chapter 4, was in providing an interval estimate for the sample median. Because the median or 50th percentile is in the center of the sample, virtually every element of the sample contributes to its determination. As we move out into the tails of a distribution, to determine the 20th percentile or the 90th, fewer and fewer elements of the sample are of assistance in making the estimate.

For a given size sample, bootstrap estimates of percentiles in the tails will always be less accurate than estimates of more centrally located percentiles. Similarly, bootstrap interval estimates for the variance of a distribution will always be less accurate than estimates of central location such as the mean or median because the variance depends strongly upon extreme values in the population.

One proposed remedy is the tilted bootstrap[3] in which instead of sampling each element of the original sample with equal probability, we weight the probabilities of selection so as to favor or discourage the selection of extreme values.

If we know something about the population distribution in advance—for example, if we know that the distribution is symmetric, or that it is chi-square with six degrees of freedom—then we may be able to take advantage of a parametric or semiparametric bootstrap as described in Chapter 4. Recognize that in doing so, you run the risk of introducing error through an inappropriate choice of parametric framework.

Problems due to the discreteness of the bootstrap statistic are usually evident from plots of bootstrap output. They can be addressed using a smooth bootstrap as described in Davison and Hinkley [1997, Section 3.4].

BAYESIAN METHODOLOGY

Since being communicated to the Royal Society in 1763,[4] Bayes' Theorem has exerted a near fatal attraction on those exposed to it.[5] Much as a bell placed on the cat would magically resolve so many of the problems of the average house mouse, Bayes' straightforward, easily grasped mathematical formula would appear to provide the long-awaited basis for a robotic judge free of human prejudice.

On the plus side, Bayes' Theorem offers three main advantages:

---

[3] See, for example, Hinkley and Shi [1989] and Phipps [1997].
[4] *Philos. Tran.* 1763; 53:376–398. Reproduced in *Biometrika* 1958; 45: 293–315.
[5] The interested reader is directed to Keynes [1921] and Redmayne [1998] for some accounts.

1. It simplifies the combination of a variety of different kinds of evidence, lab tests, animal experiments, and clinical trials, and it serves as an effective aid to decision making.
2. It permits evaluating evidence in favor of a null hypothesis. And with very large samples, a null hypothesis is not automatically rejected.
3. It provides flexibility during the conduct of an experiment; sample sizes can be modified, measuring devices altered, subject populations changed, and endpoints redefined.

Suppose we have in hand a set of evidence $E = \{E_1, E_2, \ldots, E_n\}$ and thus have determined the conditional probability $\Pr\{A \mid E\}$ that some event A is true. "A" might be the event that O.J. killed his ex-wife, that the Captain of the Valdez behaved recklessly, or some other incident whose truth or falsehood we wish to establish. An additional piece of evidence E_{n+1} now comes to light. Bayes' Theorem tell us that

$$\Pr\{A \mid E_1, \ldots, E_n, E_{n+1}\} = \frac{\Pr\{E_{n+1} \mid A\} \Pr\{A \mid E_1, \ldots, E_n\}}{\Pr\{E_{n+1} \mid A\} \Pr\{A \mid E_1, \ldots, E_n\} + \Pr\{E_{n+1} \mid {\sim}A\} \Pr\{{\sim}A \mid E_1, \ldots, E_n\}}$$

where $\sim A$, read not A, is the event that A did not occur. Recall that $\Pr\{A\} + \Pr\{{\sim}A\} = 1$. $\Pr\{A \mid E_1, \ldots, E_n\}$ is the *prior* probability of A, and $\Pr\{A \mid E_1, \ldots, E_n, E_{n+1}\}$ the *posterior* probability of A once the item of evidence E_{n+1} is in hand. Gather sufficient evidence and we shall have an automatic verdict.

The problem with the application of Bayes' Theorem in practice comes at the beginning when we have no evidence in hand, and $n = 0$. What is the prior probability of A then?

Applications in the Courtroom[6]

Bayes' Theorem has seen little use in criminal trials as ultimately the theorem relies on unproven estimates rather than known facts.[7] Tribe [1971] states several objections including the argument that a jury might actually use the evidence twice, once in its initial assessment of guilt—that is, to determine a prior probability—and a second time when the jury applies Bayes' Theorem. A further objection to the theorem's application is that if a man is innocent until proven guilty, the prior probability of his guilt must be zero; by Bayes' Theorem the posterior probability of his

---

[6] The majority of this section is reprinted with permission from *Applying Statistics in the Courtroom*, by Phillip Good, Copyright 2001 by CRC Press, Inc.

[7] See, for example, People v. Collins, 68 Cal .2d 319, 36 ALR3d 1176 (1968).

guilt would be zero also, rendering a trial unnecessary. The courts of several states have remained unmoved by this argument.[8]

In State v. Spann,[9] showing the defendant had fathered the victim's child was key to establishing a charge of sexual assault. The State's expert testified that only 1% of the presumed relevant population of possible fathers had the type of blood and tissue that the father had and, further, that the defendant was included within that 1%. In other words, 99% of the male population at large was excluded. Next, she used Bayes' Theorem to show that the defendant had a posterior probability of fathering the victim's child of 96.5%.

> "The expert testifying that the probability of defendant's paternity was 96.5% knew absolutely nothing about the facts of the case other than those revealed by blood and tissues tests of defendant, the victim, and the child . . ."[10]

> "In calculating a final probability of paternity percentage, the expert relied in part on this 99% probability of exclusion. She also relied on an assumption of a 50% prior probability that defendant was the father. This assumption [was] not based on her knowledge of any evidence whatsoever in this case . . . [she stated] "everything is equal . . . he may or may not be the father of the child."[11]

> "Was the expert's opinion valid even if the jury disagreed with the assumption of .5 [50%]? If the jury concluded that the prior probability is .4 or .6, for example, the testimony gave them no idea of the consequences, no knowledge of what the impact (of such a change in the prior probability) would be on the formula that led to the ultimate opinion of the probability of paternity."[12]

> ". . . [T]he expert's testimony should be required to include an explanation to the jury of what the probability of paternity would be for a varying range of such prior probabilities, running for example, from .1 to .9."[13]

---

[8] See, for example, Davis v. State, 476 N.E.2d 127 (Ind. App. 1985) and Griffith v. State of Texas, 976 S.W.2d 241 (1998).
[9] 130 N.J. 484 (1993)
[10] Id. 489.
[11] Id. 492.
[12] Id. 498.
[13] Id. 499.

In other words, Bayes' Theorem might prove applicable if regardless of the form of the a priori distribution, one came to more or less the same conclusion.

Courts in California,[14] Illinois, Massachusetts,[15] Utah,[16] and Virginia[17] also have challenged the use of the 50–50 assumption. In State v. Jackson,[18] the expert did include a range of prior probabilities in her testimony, but the court ruled the trial judge had erred in allowing the expert to testify as to the conclusions of Bayes' Theorem in stating a conclusion, that the defendant was 'probably' the father of the victim's child.

In Cole v. Cole,[19] a civil action, the Court rejected the admission of an expert's testimony of a high probability of paternity derived via Bayes' formula because there was strong evidence the defendant was sterile as a result of a vasectomy.

> *"The source of much controversy is the statistical formula generally used to calculate the provability of paternity: the Bayes Theorem. . . . Briefly, the Bayes Theorem shows how new statistical information alters a previously established probability. . . . When a laboratory uses the Bayes Theorem to calculate a probability of paternity it must first calculate a 'prior probability of paternity'. . . . This prior probability usually has no connection to the case at hand. Sometimes it reflects the previous success of the laboratory at excluding false fathers. Traditionally, laboratories use the figure 50% which may or may not be appropriate in a given case."*

> *"Critics suggest that this prior probability should take into account the circumstances of the particular case. For example if the woman has accused three men of fathering her child or if there are reasons to doubt her credibility, or if there is evidence that the husband is infertile, as in the present case, then the prior probability should be reduced to less than 50%."[20]*

> *The question remains as to what value to assign the prior probability. And whether absent sufficient knowledge to pin down the prior probability with any accuracy we can make use of Bayes' Theorem at all. At trial, an expert called by the prosecution in*

---

[14] State v. Jackson, 320 NC 452, 358 S.E.2d 679 (1987).

[15] Commonwealth v. Beausoleil, 397 Mass. 206 (1986).

[16] Kofford v. Flora 744 P.2d 1343, 1351-2 (1987).

[17] Bridgeman v. Commonwealth, 3 Va. App 523 (1986).

[18] 320 N.C. 452 (1987).

[19] 74 N.C. App. 247, aff'd. 314 N.C. 660 (1985).

[20] Id. 328.

Plemel v. Walter[21] *used Bayes' Theorem to derive the probability of paternity.*

"*If the paternity index or its equivalents are presented as the probability of paternity, this amounts to an unstated assumption of a prior probability of 50 percent." ". . . the paternity index will equal the probability of paternity only when the other evidence in this case establishes prior odds of paternity of exactly one.*"[22]

"*. . . the expert is unqualified to state that any single figure is the accused's 'probability of paternity.' As noted above, such a statement requires an estimation of the strength of other evidence presented in the case (i.e., an estimation of the 'prior the probability of paternity'), an estimation that the expert is no better position to make than the trier of fact.*"[23]

"*Studies in Poland and New York City have suggested that this assumption [a 50 percent prior probability] favors the putative father because in an estimated 60 to 70 percent of paternity cases the mother's accusation of paternity is correct. Of course, the purpose of paternity litigation is to determine whether the mother's accusation is correct and for that reason it would be both unfair and improper to apply the assumption in any particular case.*"[24]

A remedy proposed by the Court is of interest to us:

"*If the expert testifies to the defendant' paternity index or a substantially equivalent statistic, the expert must, if requested, calculate the probability that the defendant is the father by using more than a single assumption about the strength of the other evidence in the case. . . . If the expert uses various assumptions and makes these assumptions known, the fact finder's attention will be directed to the other evidence in the case, and it will not be misled into adopting the expert's assumption as to the correct weight to be assigned the other evidence. The expert should present calculations based on assumed prior probabilities of 0, 10, 20, . . . , 90 and 100 percent.*"[25]

---

[21] 303 Or. 262 (1987).

[22] Id. 272.

[23] Id. 275.

[24] Id. 276, fn 9.

[25] Id. 279. See also Kaye [1988].

The courts of many other states have followed *Plemmel.* "The better practice may be for the expert to testify to a range of prior probabilities, such as 10, 50 and 90 percent, and allow the trier of fact to determine which to use."[26]

Applications to Experiments and Clinical Trials

Outside the courtroom, where the rules of evidence are less rigorous, we have much greater latitude in the adoption of *a prior* distributions for the unknown parameter(s). Two approaches are common:

> 1. Adopting some synthetic distribution—a normal or a Beta.
>
> 2. Using subjective probabilities.

The synthetic approach, though common among the more computational, is difficult to justify. The theoretical basis for an observation having a normal distribution is well known—the observation will be the sum of a large number of factors, each of which makes only a minute contribution to the total. But could such a description be applicable to a population parameter?

Here is an example of this approach taken from a report by D. A. Berry[27]: "A study reported by Freireich et al.[28] was designed to evaluate the effectiveness of a chemotherapeutic agent 6-mercaptopurine (6-MP) for the treatment of acute leukemia. Patients were randomized to therapy in pairs. Let p be the population proportion of pairs in which the 6-MP patient stays in remission longer than the placebo patient. (To distinguish probability p from a probability distribution concerning p, I will call it a population proportion or a propensity.) The null hypothesis H_0 is $p = 1/2$: no effect of 6-MP. Let H_1 stand for the alternative hypothesis that $p > 1/2$. There were 21 pairs of patients in the study, and 18 of them favored 6-MP."

"Suppose that the prior probability of the null hypothesis is 70 percent and that the remaining probability of 30 percent is on the interval $(0,1)$ uniformly. . . . So under the alternative hypothesis H_1, p has a uniform$(0,1)$ distribution. This is a mixture prior in the sense that it is 70 percent discrete and 30 percent continuous."

---

[26] County of El Dorado v. Misura, 33 Cal. App. 4th 73 (1995) citing Plemel, supra, at p. 1219; Peterson (1982 at p. 691, fn. 74), Paternity of M.J.B., 144 Wis.2d 638, 643; State v. Jackson, 320 N.C.452, 455 (1987), and Kammer v. Young, 73 Md. App. 565, 571 (1988). See also State v. Spann, 130 N.J. 484 at p. 499 (1993).

[27] The full report titled "Using a Bayesian Approach in Medical Device Development" may be obtained from Donald A. Berry at the Institute of Statistics & Decision Sciences and Comprehensive Cancer Center, Duke University, Durham, NC 27708.

[28] *Blood* 1963; 21:699–716.

"The uniform(0,1) distribution is also the beta(1,1) distribution. Updating the beta(a,b) distribution after s successes and f failures is easy, namely, the new distribution is beta($a + s, b + f$). So for $s = 18$ and $f = 3$, the posterior distribution under H_1 is beta(19,4)."

The subjective approach places an added burden on the experimenter. As always, she needs to specify each of the following:

- Maximum acceptable frequency of Type I errors (that is, the significance level)
- Alternative hypotheses of interest
- Power desired against each alternative
- Losses associated with Type I and Type II errors

With the Bayesian approach, she must also provide *a priori* probabilities.

Arguing in favor of the use of subjective probabilities is that they permit incorporation of expert judgment in a formal way into inferences and decision-making. Arguing against them in the words of the late Edward Barankin, "How are you planning to get these values—beat them out of the researcher?" More appealing, if perhaps no more successful, approaches are described by Good [1950] and Kadane et al. [1980].

Bayes' Factor

An approach that allows us to take advantage of the opportunities Bayes' Theorem provides while avoiding its limitations and the objections raised in the courts is through the use of the minimum Bayes' factor introduced by Edwards et al. [1963].

The Bayes factor is a measure of the degree to which the data from a study moves us from our initial position. Let B denote the odds we put on the primary hypothesis before we examine the data, and let A be the odds we assign after seeing the data; the Bayes factor is defined as A/B.

If the Bayes factor is equal to 1/10th, it means that the study results have decreased the relative odds assigned to the primary hypothesis by tenfold. For example, suppose the probability of the primary hypothesis with respect to the alternate hypothesis was high to begin with, say 9 to 1. A tenfold decrease would mean a change to odds of 9 to 10, a probability of 47%. A further independent study with a Bayes factor of 1/10th would mean a change to a posteriori odds of 9 to 100, less than 9%.

The minimum Bayes factor is calculated from the same information used to determine the p value, and it can easily be derived from standard analytic results. In the words of Goodman [2001], "If a statistical test is based on a Gaussian approximation, the strongest Bayes factor against the null hypothesis is $\exp(-Z^2/2)$, where Z is the number of standard errors from the null value. If the log-likelihood of a model is reported, the minimum Bayes factor is simply the exponential of the difference between the log-likelihoods of two competing models (i.e., the ratio of their maximum likelihoods)."

The minimum Bayes factor does not involve a *specific* prior probability distribution, rather, it is a global minimum over *all* prior distributions. Bayarri and Berger [1998] and Berger and Sellke [1987]] provide a simple formula for the minimum Bayes factor in the situation where the prior probability distribution is symmetric and descending around the null value. This is $-\exp p \ln(p)$, where p is the fixed-sample-size p value.

As Goodman [2001] notes, "even the strongest evidence against the null hypothesis does not lower its odds as much as the p-value magnitude might lead people to believe. More importantly, the minimum Bayes factor makes it clear that we cannot estimate the credibility of the null hypothesis without considering evidence outside the study."

For example, while a p value of 0.01 is usually termed "highly significant," it actually represents evidence for the primary hypothesis of somewhere between $1/25$ and $1/8$.[29] Put another way, the relative odds of the primary hypothesis versus any alternative given a p value of 0.01 are at most 8–25 times lower than they were before the study. If one is going to claim that a hypothesis is highly unlikely (e.g., less than 5%), one must already have evidence outside the study that the prior probability of the hypothesis is no greater than 60%. Conversely, even weak evidence in support of a highly plausible relationship may be enough for an author to make a convincing case.

Two Caveats

1. Bayesian methods cannot be used in support of after-the-fact-hypotheses because, by definition, an after-the-fact hypothesis has zero *a priori* probability and, thus, by Bayes' rule, zero *a posteriori* probability.

2. One hypothesis proving of greater predictive value than another in a given instance may be suggestive but is far from definitive in the absence of collateral evidence and proof of causal mechanisms. See, for example, Hodges [1987].

When Using Bayesian Methods

Do not use an arbitrary prior.

Never report a p value.

Incorporate potential losses in the decision.

Report the Bayes factor.

---

[29] See Table B.1, Goodman [2001].

META-ANALYSIS

"Meta-analysis should be viewed as an observational study of the evidence. The steps involved are similar to any other research undertaking: formulation of the problem to be addressed, collection and analysis of the data, and reporting of the results. Researchers should write in advance a detailed research protocol that clearly states the objectives, the hypotheses to be tested, the subgroups of interest, and the proposed methods and criteria for identifying and selecting relevant studies and extracting and analysing information" (Egger, Smith, and Phillips, 1997).[30]

Too many studies end with inconclusive results because of the relatively small number of observations that were made. The researcher can't quite reject the null hypothesis, but isn't quite ready to embrace the null hypothesis, either. As we saw in Chapter 1, a post hoc subgroup analysis can suggest an additional relationship, but the relationship cannot be subject to statistical test in the absence of additional data.

Meta-analysis is a set of techniques that allow us to combine the results of a series of small trials and observational studies. With the appropriate meta-analysis, we can, in theory, obtain more precise estimates of main effects, test *a priori* hypotheses about subgroups, and determine the number of observations needed for large-scale randomized trials.

By putting together all available data, meta-analyses are also better placed than individual trials to answer questions about whether an overall study result varies among subgroups—for example, among men and women, older and younger patients, or subjects with different degrees of severity of disease.

In performing a meta-analysis, we need to distinguish between observational studies and randomized trials.

Confounding and selection bias can easily distort the findings from observational studies. Egger et al. [1998] note, "An important criterion supporting causality of associations is a dose–response relation. In occupational epidemiology the quest to show such an association can lead to very different groups of employees being compared. In a meta-analysis that examined the link between exposure to formaldehyde and cancer, funeral directors and embalmers (high exposure) were compared with anatomists and pathologists (intermediate to high exposure) and with industrial workers (low to high exposure, depending on job assignment). There is a striking deficit of deaths from lung cancer among anatomists and pathologists [standardized mortality ratio 33 (95% confidence interval 22 to 47)], which is most likely to be due to a lower prevalence of smoking among

---

[30] Reprinted with permission from the BMJ Publishing Group.

this group. In this situation few would argue that formaldehyde protects against lung cancer. In other instances, however, such selection bias may be less obvious."[31]

On the other hand, much may be gained by a careful examination of possible sources of heterogeneity between the results from observational studies.

Publication and selection bias also plague the meta-analysis of completely randomized trials. Inconclusive or negative results seldom appear in print (Götzsche, 1987; Chalmers et al., 1990; Easterbrook et al., 1991) and are unlikely even to be submitted for publication. One can't analyze what one doesn't know about.

Similarly, the decision as to which studies to incorporate can dramatically affect the results. Meta-analyses of the same issue may reach opposite conclusions, as shown by assessments of low-molecular-weight heparin in the prevention of perioperative thrombosis (Nurmohamed et al., 1992; Leizorovicz et al., 1992) and of second line antirheumatic drugs in the treatment of rheumatoid arthritis (Felson et al., 1990; Götzsche et al., 1992). Meta-analyses showing benefit of statistical significance and clinical importance have been contradicted later by large randomized trials (Egger et al., 1997).

Where there are substantial differences between the different studies incorporated in a meta-analysis (their subjects or their environments), or substantial quantitative differences in the results from the different trials, a single overall summary estimate of treatment benefit has little practical applicability (Horowitz, 1995). Any analysis that ignores this heterogeneity is clinically misleading and scientifically naive (Thompson, 1994). Heterogeneity should be scrutinized, with an attempt to explain it (Bailey, 1987; Berkey et al., 1995; Chalmers, 1991; Victor, 1995).

Bayesian Methods

Bayesian methods can be effective in meta-analyses; see, for example, Mosteller and Chalmers [1992]. In such situations the parameters of various trials are considered to be random samples from a distribution of trial parameters. The parameters of this higher-level distribution are called hyperparameters, and they also have distributions. The model is called *hierarchical*. The extent to which the various trials reinforce each other is determined by the data. If the trials are very similar, the variation of the hyperparameters will be small, and the analysis will be very close to a classical meta-analysis. If the trials do not reinforce each other, the conclusions of the hierarchical Bayesian analysis will show a very high variance in the results.

---

[31] Reprinted with permission from the BMJ Publishing Group.

A hierarchical Bayesian analysis avoids the necessity of a prior decision as to whether or not the trials can be combined; the extent of the combination is determined purely by the data. This does not come for free; in contrast to the meta-analyses discussed above, all the original data (or at least the sufficient statistics) must be available for inclusion in the hierarchical model. The Bayesian method is also vulnerable to all the selection bias issues discussed above.

Guidelines For a Meta-Analysis

- A detailed research protocol for the meta-analysis should be prepared in advance. Criteria for inclusion and statistical method employed should be documented in the materials and methods section of the subsequent report.
- Meta-analysis should be restricted to randomized controlled trials.
- Heterogeneity in the trial results should be documented and explained.
- Do not attempt to compare treatments investigated in unrelated trials. (Suppose, by way of a counterexample, that Old were given as always to low-risk patients in one set of trials, while New was given to high-risk patients in another.)
- Individual patient data, rather than published summary statistics, often are required for meaningful subgroup analyses. This is a major reason why we favor the modern trend of journals to insist that all data reported on within their pages be made available by website to all investigators.

Kepler was able to formulate his laws only because (1) Tycho Brahe had made over 30 years of precise (for the time) astronomical observations and (2) Kepler married Brahe's daughter and, thus, gained access to his data.

PERMUTATION TESTS

Permutation tests are often lauded erroneously in the literature as "assumption-free" "panaceas." Nothing could be further from the truth.

Permutation tests only yield exact significance levels if the labels on the observations are weakly exchangeable under the null hypothesis. Thus, they cannot be successfully applied to the coefficients in a multivariate regression.

On the other hand, if the observations are weakly exchangeable under the null hypothesis, then permutation tests are the method of choice for k-sample comparisons, multi-factor experimental designs, and contingency tables, whenever there are 12 or less observations in each subsample. Moreover, permutation methods can be used both to test hypotheses and to obtain interval estimates of parameters.

TO LEARN MORE

Potential flaws in the bootstrap approach are considered by Schenker [1985], Wu [1986], Diciccio and Romano [1988], Efron [1988, 1992], Knight [1989], and Gine and Zinn [1989]. Canty et al. [2000] provide a set of diagnostics for detecting and dealing with potential error sources.

Berry and Stangl [1996] include a collection of case studies in Bayesian biostatistics. Kass and Raftery [1995] discuss the problem of establishing priors along with a set of practical examples. The Bayes factor can be used as a test statistic; see Good [1992].

For more on the strengths and limitations of meta-analysis, see Egger and Smith [1997], Egger, Smith, and Phillips [1997], Smith, Egger, and Phillips [1997], Smith and Egger [1998], Gillett [2001], Gurevitch and Hedges [1993], Horowitz [1995], and Smeeth, Haines, and Ebrahim [1999]. To learn about the appropriate statistical procedures, see Adams, Gurevitch, and Rosenberg [1997], Berlin et al. [1989], and Hedges and Olkin [1985].

For practical, worked-through examples of hierarchical Bayesian analysis, see Harley and Myers [2001] and Su, Adkison, and Van Alen [2001]. Theoretical development may be found in Mosteller and Chalmers [1992] and in Carlin and Louis [1996].

The lack of access to the raw data underlying published studies is a matter of ongoing concern. See Moher et al. [1999], Eysenbach and Sa [2001], and Hutchon [2001].

Permutation methods and their applications are described in Good [2001], Manley [1997], Mielke and Berry [2001], and Pesarin [2001]. For a description of some robust permutation tests, see Lambert [1985] and Maritz [1996]. Berger [2000] reviews the pros and cons of permutation tests.

Chapter 7
Reporting Your Results

"Cut out the appropriate part of the computer output and paste it onto the draft of the paper." George Dyke (tongue in cheek) [1997].

THE FOCUS OF THIS CHAPTER IS ON WHAT to report and how to report it. Reportable elements include the experimental design and its objectives, its analysis, and the sources and amounts of missing data. Guidelines for table construction are provided. The bootstrap is proposed as an alternative to the standard error as a measure of precision. The value and limitations of p values and confidence intervals are summarized. Practical significance is distinguished from statistical significance and induction from deduction.

FUNDAMENTALS

Few experimenters fail to list number of subjects, doses administered, and dose intervals in their reports. But many fail to provide the details of power and sample size calculations. Feng et al. [2001] found that such careless investigators also report a higher proportion of nonsignificant intervention effects, indicating underpowered studies.

Too often inadequate attention is given to describing treatment allocation and the ones who got away. We consider both topics in what follows.

Treatment Allocation[1]
Allocation details should be fully described in your reports including dictated allocation versus allocation discretion, randomization, advance preparation of the allocation sequence, allocation concealment, fixed versus varying allocation proportions, restricted randomization, masking, simulta-

---

[1] This material in this section relies heavily on a personal communication from Vance W. Berger and Costas A. Christophi.

Common Errors in Statistics (and How to Avoid Them), by Phillip I. Good and James W. Hardin.
ISBN 0-471-46068-0   Copyright © 2003 John Wiley & Sons, Inc.

neous versus sequential randomization, enrollment discretion, and the possibility of intent to treat.

Allocation discretion may be available to the investigator, the patient, both, or neither (dictated allocation). Were investigators permitted to assign treatment based on patient characteristics? Could patients select their own treatment from among a given set of choices?

Was actual (not virtual, quasi-, or pseudo-) randomization employed? Was the allocation sequence predictable? (For example, patients with even accession numbers or patients with odd accession numbers receive the active treatment; the others receive the control.)

Was randomization *conventional*, that is, was the allocation sequence generated in advance of screening any patients?

Was allocation concealed prior to its being executed? As Vance W. Berger and Costas A. Christophi relate in a personal communication, "This is not itself a reportable design feature, so a claim of allocation concealment should be accompanied by specific design features. For example, one may conceal the allocation sequence; and instead of using envelopes, patient enrollment may involve calling the baseline information of the patient to be enrolled in to a central number to receive the allocation."

Was randomization restricted or unrestricted? Randomization is *unrestricted* if a patient's likelihood of receiving either treatment is independent of all previous allocations and is *restricted* otherwise. If both treatment groups must be assigned equally often, then prior allocations determine the final ones. Were the proportions also hidden?

Were treatment codes concealed until all patients had been randomized and the database locked? Were there instances of codes being revealed accidentally? Senn [1995] warns, "investigators should delude neither themselves, nor those who read their results, into believing that simply because some aspects of their trial were double-blind that therefore all the virtues of such trials apply to all their conclusions." Masking can rarely, if ever, be assured; see, also, Day [1998].

Was randomization simultaneous, block simultaneous, or *sequential*? A blocked randomization is *block simultaneous* if all patients within any given block are identified and assigned accession numbers prior to any patient in that block being treated.

And, not least, was intent to treat permitted?

Missing Data[2]

Every experiment or survey has its exceptions. You must report the raw numbers of such exceptions and, in some instances, provide additional

---

[2] Material in this section is reprinted with permission from *Manager's Guide to Design and Conduct of Clinical Trials*, Wiley, 2002.

analyses that analyze or compensate for them. Typical exceptions include the following:

Did Not Participate. Subjects who were eligible and available but did not participate in the study—this group should be broken down further into those who were approached but chose not to participate and those who were not approached. With a mail-in survey for example, we would distinguish between those whose envelopes were returned "address unknown" and those who simply did not reply.

Ineligibles. In some instances, circumstances may not permit deferring treatment until the subject's eligibility can be determined.

For example, an individual arrives at a study center in critical condition; the study protocol calls for a series of tests, the results of which may not be back for several days; but in the opinion of the examining physician, treatment must begin immediately. The patient is randomized to treatment, and only later is it determined that the patient is ineligible.

The solution is to present two forms of the final analysis: one incorporating all patients, the other limited to those who were actually eligible.

Withdrawals. Subjects who enrolled in the study but did not complete it including both dropouts and noncompliant patients. These patients might be subdivided further based on the point in the study at which they dropped out.

At issue is whether such withdrawals were treatment related or not. For example, the gastrointestinal side effects associated with erythromycin are such that many patients (including both authors) may refuse to continue with the drug. Traditional statistical methods are not applicable when withdrawals are treatment related.

Crossovers. If the design provided for intent-to-treat, a noncompliant patient may still continue in the study after being reassigned to an alternate treatment. Two sets of results should be reported: the first for all patients who completed the trials (retaining their original treatment assignments for the purpose of analysis), the second restricted to the smaller number of patients who persisted in the treatment groups to which they were originally assigned.

Missing Data. Missing data is common, expensive, and preventable in many instances.

The primary endpoint of a recent clinical study of various cardiovascular techniques was based on the analysis of follow-up angiograms. Although more than 750 patients were enrolled in the study, only 523 had the necessary angiograms. Almost one-third of the monies spent on the trials had been wasted. This result is not atypical. Capaldi and Patterson

[1987] uncovered an average attrition rate of 47% in studies lasting 4 to 10 years.

You need to analyze the data to ensure that the proportions of missing observations are the same in all treatment groups. Again, traditional statistical methods are applicable only if missing data are not treatment related.

Deaths and disabling accidents and diseases, whether or not directly related to the condition being treated, are common in long-term trials in the elderly and high-risk populations. Or individuals are simply lost to sight ("no forwarding address") in highly mobile populations.

Lang and Secic [1997, p. 22] suggest a chart such as that depicted in Figure 3.1 as the most effective way to communicate all the information regarding missing data. Censored and off-scale measurements should be described separately and their numbers indicated in the corresponding tables.

TABLES

Is text, a table, or a graph the best means of presenting results? Dyke [1997] would argue, "Tables with appropriate marginal means are often the best method of presenting results, occasionally replaced (or supplemented) by diagrams, usually graphs or histograms." Van Belle [2002] warns that aberrant values often can be more apparent in graphical form. Arguing in favor of the use of ActivStats® for exploratory analysis is that one can so easily go back and forth from viewing the table to viewing the graph.

A sentence structure should be used for displaying two to five numbers, as in "The blood type of the population of the United States is approximately 45% O, 40% A, 11% B, and 4% AB."[3] Note that the blood types are ordered by frequency.

Marginal means may be omitted only if they have already appeared in other tables.[4] Sample sizes should always be specified.

Among our own worst offenses is the failure to follow van Belle's advice to "Use the table heading to convey critical information. Do not stint. The more informative the heading, the better the table."[5]

Consider adding a row (or column, or both) of contrasts; "for example, if the table has only two rows we could add a row of differences, row 1 minus row 2: if there are more than two rows, some other contrast might be useful, perhaps 'mean haploid minus mean diploid', or 'linear component of effect of N-fertilizer'."[6] Indicate the variability of these contrasts.

---

[3] van Belle [2002, p. 154].
[4] Dyke [1997]. Reprinted with permission from Elsevier Science.
[5] van Belle [2002, p. 154].
[6] Dyke [1997]. Reprinted with permission from Elsevier Science.

Tables dealing with two-factor arrays are straightforward, provided that confidence limits, least standard deviations, and standard errors are clearly associated with the correct set of figures. Tables involving three or more factors are not always immediately clear to the reader and are best avoided.

Are the results expressed in appropriate units? For example, are parts per thousand more natural in a specific case than percentages? Have we rounded off to the correct degree of precision, taking account of what we know about the variability of the results and considering whether they will be used by the reader, perhaps by multiplying by a constant factor or by another variate—for example, % dry matter?

Dyke [1997] also advises us that "Residuals should be tabulated and presented as part of routine analysis; any [statistical] package that does not offer this option was probably produced by someone out of touch with research workers, certainly with those working with field crops." Best of all is a display of residuals aligned in rows and columns as the plots were aligned in the field.

A table of residuals (or tables, if there are several strata) can alert us to the presence of outliers and may also reveal patterns in the data not considered previously.

STANDARD ERROR

One of the most egregious errors in statistics—one encouraged, if not insisted upon by the editors of journals in the biological and social sciences—is the use of the notation "mean ± standard error" to report the results of a set of observations.

Presumably, the editors of these journals (and the reviewers they select) have three objectives in mind: To communicate some idea of

1. The "correct" result
2. The precision of the estimate of the correct result
3. The dispersion of the distribution from which the observations were drawn

Let's see to what degree any or all of these objectives might be realized in real life by the editor's choice.

For small samples of three to five observations, summary statistics are virtually meaningless; reproduce the actual observations; this is easier to do and more informative.

For many variables, regardless of sample size, the arithmetic mean can be very misleading. For example, the mean income in most countries is far in excess of the median income or 50th percentile to which most of us can relate. When the arithmetic mean is meaningful, it is usually equal to or close to the median. Consider reporting the median in the first place.

The *geometric mean* is more appropriate than the arithmetic in three sets of circumstances:

1. **When losses or gains can best be expressed as a percentage rather than a fixed value.**
2. **When rapid growth is involved as in the development of a bacterial or viral population.**
3. **When the data span several orders of magnitude, as with the concentration of pollutants.**

Because bacterial populations can double in number in only a few hours, many government health regulations utilize the geometric rather than the arithmetic mean.[7] A number of other government regulations also use it, though the sample median would be far more appropriate.[8]

Whether you report a mean or a median, be sure to report only a sensible number of decimal places. Most statistical packages can give you 9 or 10. Don't use them. If your observations were to the nearest integer, your report on the mean should include only a single decimal place. For guides to the appropriate number of digits, see Ehrenberg [1977]; for percentages, see van Belle [2002, Table 7.4].

The standard error is a useful measure of population dispersion *if* the observations come from a normal or Gaussian distribution. If the observations are normally distributed as in the bell-shaped curve depicted in Figure 7.1, then in 95% of the samples we would expect the sample mean to lie within two standard errors of the mean of our original sample.

But if the observations come from a nonsymmetric distribution like an exponential or a Poisson, or a truncated distribution like the uniform, or a mixture of populations, we cannot draw any such inference.

Recall that the standard error equals the standard deviation divided by the square root of the sample size, $SE = \dfrac{\sum (x_i - \bar{x})^2}{\sqrt{n(n-1)}}$. Because the standard error depends on the squares of individual observations, it is particularly sensitive to outliers. A few extra large observations will have a dramatic impact on its value.

If you can't be sure your observations come from a normal distribution, then consider reporting your results either in the form of a histogram (as in Figure 7.2) or in a box and whiskers plot (Figure 7.3). See also Lang and Secic [1997, p. 50].

---

[7] See, for example, 40 CFR part 131, 62 Fed. Reg. 23004 at 23008 (28 April 1997).
[8] Examples include 62 Fed. Reg. 45966 at 45983 (concerning the length of a hospital stay) and 62 Fed. Reg. 45116 at 45120 (concerning sulfur dioxide emissions).

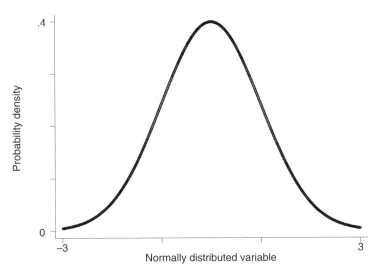

FIGURE 7.1 Bell-Shaped Symmetric Curve of a Normal Distribution.

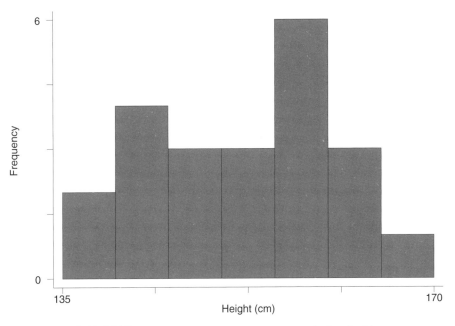

FIGURE 7.2 Histogram of Heights in a Sixth-Grade Class.

If your objective is to report the precision of your estimate of the mean or median, then the standard error may be meaningful providing the mean of your observations is normally distributed.

The good news is that the sample mean often will have a normal distribution even when the observations themselves do not come from a

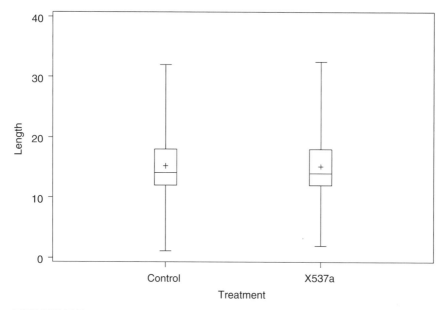

FIGURE 7.3 **Box and Whiskers Plot.** The box encompasses the middle 50% of each sample while the "whiskers" lead to the smallest and largest values. The line through the box is the median of the sample; that is, 50% of the sample is larger than this value, while 50% is smaller. The plus sign indicates the sample mean. Note that the mean is shifted in the direction of a small number of very large values.

normal distribution. This is because the sum of a large number of random variables each of which makes only a small contribution to the total is a normally distributed random variable.[9] And in a sample mean based on n observations, each contributes only $1/n$th the total. How close the fit is to a normal will depend upon the size of the sample and the distribution from which the observations are drawn.

The distribution of a uniform random number $U[0,1]$ is a far cry from the bell-shaped curve of Figure 7.1. Only values between 0 and 1 have a positive probability, and in stark contrast to the normal distribution, no range of values between zero and one is more likely than another of the same length. The only element the uniform and the normal distributions have in common is their symmetry about the population mean. Yet to obtain normally distributed random numbers for use in simulations a frequently employed technique is to generate 12 uniformly distributed random numbers and then take their average.

---

[9] This result is generally referred to as the Central Limit Theorem. Formal proof can be found in a number of texts including Feller [1966, p. 253].

| | | | | |||| | | | | | |||| | | | || | ||| |
142.25   Medians of bootstrap samples   158.25

FIGURE 7.4 Rugplot of 50 Bootstrap Medians Derived from a Sample of Sixth Grader's Heights.

Apparently, 12 is a large enough number for a sample mean to be normally distributed when the variables come from a uniform distribution. But if you take a smaller sample of observations from a $U[0,1]$ population, the distribution of its mean would look less like a bell-shaped curve.

A loose rule of thumb is that the mean of a sample of 8 to 25 observations will have a distribution that is close enough to the normal for the standard error to be meaningful. The more nonsymmetric the original distribution, the larger the sample size required. At least 25 observations are needed for a binomial distribution with $p = 0.1$.

Even the mean of observations taken from a mixture of distributions (males and females, tall Zulu and short Bantu)—visualize a distribution curve resembling a camel with multiple humps—will have a normal distribution if the sample size is large enough. Of course, this mean (or even the median) conceals the fact that the sample was taken from a mixture of distributions.

If the underlying distribution is not symmetric, the use of the ± SE notation can be deceptive because it suggests a nonexistent symmetry. For samples from nonsymmetric distributions of size 6 or less, tabulate the minimum, the median, and the maximum. For samples of size 7 and up, consider using a box and whiskers plot as in Figure 7.3. For samples of size 16 and up, the bootstrap (described in Chapters 4 and 5) may provide the answer you need.

As in Chapters 4 and 5, we would treat the original sample as a stand-in for the population and resample from it repeatedly, 1000 times or so, with replacement, computing the sample statistic each time to obtain a distribution similar to that depicted in Figure 7.4. To provide an interpretation compatible with that given the standard error when used with a sample from a normally distributed population, we would want to report the values of the 16th and 84th percentiles of the bootstrap distribution along with the sample statistic.

When the estimator is other than the mean, we cannot count on the Central Limit Theorem to ensure a symmetric sampling distribution. We recommend you use the bootstrap whenever you report an estimate of a ratio or dispersion.

If you possess some prior knowledge of the shape of the population distribution, you should take advantage of that knowledge by using a parametric bootstrap (see Chapter 4). The parametric bootstrap is particularly recommended for use in determining the precision of percentiles in the tails (P_{20}, P_{10}, P_{90}, and so forth).

p VALUES

Before interpreting and commenting on p values, it's well to remember that in contrast to the significance level, the p value is a random variable that varies from sample to sample. There may be highly significant differences between two populations and yet the samples taken from those populations and the resulting p value may not reveal that difference. Consequently, it is not appropriate for us to compare the p values from two distinct experiments, or from tests on two variables measured in the same experiment, and declare that one is more significant than the other.

If in advance of examining the data we agree that we will reject the hypothesis if the p value is less than 5%, then our significance level is 5%. Whether our p value proves to be 4.9% or 1% or 0.001%, we will come to the same conclusion. One set of results is not more significant than another; it is only that the difference we uncovered was measurably more extreme in one set of samples than in another.

p values need not reflect the strength of a relationship. Duggan and Dean [1968] reviewed 45 articles that had appeared in sociology journals between 1955 and 1965 in which the chi-square statistic and distribution had been employed in the analysis of 3×3 contingency tables and compared the resulting p values with association as measured by Goodman and Kruskal's gamma. Table 7.1 summarizes their findings.

p values derived from tables are often crude approximations, particularly for small samples and tests based on a specific distribution. They and the stated significance level of our test may well be in error.

The vast majority of p values produced by parametric tests based on the normal distribution are approximations. If the data are "almost" normal, the associated p values will be almost correct. As noted in Chapter 6, the stated significance values for Student's t are very close to exact. Of course a stated p value of 4.9% might really prove to be 5.1% in practice. The significance values associated with the F statistic can be completely inaccurate for non-normal data (1% rather than 10%). And the p values derived from

TABLE 7.1 p-Value and Association

p-value	gamma		
	<.30	.30–.70	>.70
<.01	8	11	5
.05	7	0	0
>.10	8	0	0

the chi-square distribution for use with contingency tables also can be off by an order of magnitude.

The good news is that there exists a class of tests, the permutation tests described in Chapter 5, for which the significance levels are exact if the observations are independent and identically distributed under the null hypothesis or their labels are otherwise exchangeable.

CONFIDENCE INTERVALS

If p values are misleading, what are we to use in their place? Jones [1955, p. 407] was among the first to suggest that "an investigator would be misled less frequently and would be more likely to obtain the information he seeks were he to formulate his experimental problems in terms of the estimation of population parameters, with the establishment of confidence intervals about the estimated values, rather than in terms of a null hypothesis against all possible alternatives." See also Gardner and Altman [1996] and Poole [2001].

Confidence intervals can be derived from the rejection regions of our hypothesis tests, whether the latter are based on parametric or nonparametric methods. Suppose $A(\theta')$ is a $1 - \alpha$ level acceptance region for testing the hypothesis $\theta = \theta'$; that is, we accept the hypothesis if our test statistic T belongs to the acceptance region $A(\theta')$ and reject it otherwise. Let $S(X)$ consist of all the parameter values $\theta^\star$ for which $T[X]$ belongs to the acceptance region $A(\theta^\star)$. Then $S(X)$ is a $1 - \alpha$ level confidence interval for θ based on the set of observations $X = \{x_1, x_2, \ldots, x_n\}$.

The probability that $S(X)$ includes θ_0 when $\theta = \theta_0$ is equal to the probability that $T(X)$ belongs to the acceptance region of θ_0 and is greater than or equal to α.

As our confidence $1 - \alpha$ increases, from 90% to 95%, for example, the width of the resulting confidence interval increases. Thus, a 95% confidence interval is wider than a 90% confidence interval.

By the same process, the rejection regions of our hypothesis tests can be derived from confidence intervals. Suppose our hypothesis is that the odds ratio for a 2×2 contingency table is 1. Then we would accept this null hypothesis if and only if our confidence interval for the odds ratio includes the value 1.

A common error is to misinterpret the confidence interval as a statement about the unknown parameter. It is not true that the probability that a parameter is included in a 95% confidence interval is 95%. What is true is that if we derive a large number of 95% confidence intervals, we can expect the true value of the parameter to be included in the computed intervals 95% of the time. (That is, the true values will be included *if* the assumptions on which the tests and confidence intervals are based are satisfied 100% of the time.) Like the p value, the upper and lower confidence

IMPORTANT TERMS

Acceptance Region, $A(\theta_0)$. Set of values of the statistic $T[X]$ for which we would accept the hypothesis $H: \theta = \theta_0$. Its complement is called the rejection region.

Confidence Region, $S(X)$. Also referred to as a confidence interval (for a single parameter) or a confidence ellipse (for multiple parameters). Set of values of the parameter θ for which given the set of observations $X = \{x_1, x_2, \ldots, x_n\}$ and the statistic $T[X]$ we would accept the corresponding responding hypothesis.

limits of a particular confidence interval are random variables because they depend upon the sample that is drawn.

Confidence intervals can be used both to evaluate and to report on the precision of estimates (see Chapter 4) and the significance of hypothesis tests (see Chapter 5). The probability the interval covers the true value of the parameter of interest and the method used to derive the interval must also be reported.

In interpreting a confidence interval based on a test of significance, it is essential to realize that the center of the interval is no more likely than any other value, and the confidence to be placed in the interval is no greater than the confidence we have in the experimental design and statistical test it is based upon. (As always, GIGO.)

Multiple Tests

Whether we report p values or confidence intervals, we need to correct for multiple tests as described in Chapter 5. The correction should be based on the number of tests we *perform*, which in most cases will be larger than the number on which we report.

RECOGNIZING AND REPORTING BIASES

Very few studies can avoid bias at some point in sample selection, study conduct, and results interpretation. We focus on the wrong endpoints; participants and co-investigators see through our blinding schemes; the effects of neglected and unobserved confounding factors overwhelm and outweigh the effects of our variables of interest. With careful and prolonged planning, we may reduce or eliminate many potential sources of bias, but seldom will we be able to eliminate all of them. Accept bias as inevitable and then endeavor to recognize and report all exceptions that do slip through the cracks.

Most biases occur during data collection, often as a result of taking observations from an unrepresentative subset of the population rather than from the population as a whole. The example of the erroneous forecast of Landon over Roosevelt was cited in Chapter 3. In Chapter 5, we considered a study that was flawed because of a failure to include planes that did *not* return from combat.

When analyzing extended time series in seismological and neurological and investigations, investigators typically select specific cuts (a set of consecutive observations in time) for detailed analysis, rather than trying to examine all the data (a near impossibility). Not surprisingly, such "cuts" usually possess one or more intriguing features not to be found in run-of-the-mill samples. Too often, theories evolve from these very biased selections. We expand on this point in Chapter 9 in our discussion of the limitations on the range over which a model may be applied.

Limitations in the measuring instrument such as censoring at either end of the scale can result in biased estimates. Current methods of estimating cloud optical depth from satellite measurements produce biased results that depend strongly on satellite viewing geometry. In this and in similar cases in the physical sciences, absent the appropriate nomograms and conversion tables, interpretation is impossible.

Over- and underreporting plague meta-analysis (discussed in Chapter 6). Positive results are reported for publication, negative findings are suppressed or ignored. Medical records are known to underemphasize conditions (such as arthritis) for which there is no immediately available treatment while overemphasizing the disease of the day. (See, for example, Callaham et al. [1998].)

Collaboration between the statistician and the domain expert is essential if all sources of bias are to be detected and corrected for, because many biases are specific to a given application area. In the measurement of price indices, for example, the three principal sources are substitution bias, quality change bias, and new product bias.[10]

Two distinct kinds of statistical bias effects arise with astronomical distance indicators (DIs), depending on the method used.[11] "In one approach, the redshifts of objects whose DI-inferred distances are within a narrow range of some value d are averaged. Subtracting d from the resulting mean redshift yields a peculiar velocity estimate; dividing the mean redshift by d gives an estimate of the parameter of interest. These estimates will be biased because the distance estimate d itself is biased and is not the mean true distance of the objects in question."

"This effect is called homogeneous Malmquist bias. It tells us that, typically, objects lie further away than their DI-inferred distances. The physical cause is more objects 'scatter in' from larger true distances (where there is more volume) than 'scatter out' from smaller ones."

---

[10] Otmar Issing in a speech at the CEPR/ECB Workshop on issues in the measurement of price indices, Frankfurt am Main, 16 November 2001.

[11] These next paragraphs are taken with minor changes from Willick [1999, Section 9].

"A second sort of bias comes into play because some galaxies are too faint or small to be in the sample; in effect, the large-distance tail of $P(d/r)$ is cut off. It follows that the typical inferred distances are smaller than those expected at a given true distance r. As a result, the peculiar velocity model that allows true distance to be estimated as a function of redshift is tricked into returning shorter distances. This bias goes in the same sense as Malmquist bias, but is fundamentally different." It results not from volume/density effects, but from the same sort of sample selection effects that were discussed earlier in this section.

Selection bias can be minimized by working in the "inverse direction." Rather than trying to predict absolute magnitude (Y) given a value of the velocity width parameter (X), instead one fits a line by regressing the widths X on the magnitudes Y.

Finally, bias can result from grouping or averaging data. Bias if group randomized trials are analyzed without correcting for cluster effects was reported by Feng et al. [1996]; see Chapter 5. The use of averaged rather than end-of-period data in financial research results in biased estimates of the variance, covariance, and autocorrelation of the first- as well as higher-order changes. Such biases can be both time varying and persistent (Wilson, Jones, and Lundstrum, 2001).

REPORTING POWER

Statisticians are routinely forced to guess at the values of population parameters in order to make the power calculations needed to determine sample size. Once the data are in hand, it's tempting to redo these same power calculations. Don't. Post hoc calculations invariably inflate the actual power of the test (Zumbo and Hubley, 1998).

Post hoc power calculations can be of value in designing follow-up studies, but should not be used in reports.

DRAWING CONCLUSIONS

Found data (nonrandom samples) can be very useful in suggesting models and hypotheses for further exploration. But without a randomized study, formal inferential statistical analyses are not supported (Greenland, 1990; Rothman, 1990b). The concepts of significance level, power, p value, and confidence interval apply only to data that have arisen from carefully designed and executed experiments and surveys.

A vast literature has grown up around the unease researchers feel in placing too much reliance on p values. Examples include Selvin [1957], Yoccuz [1991], Badrick and Flatman [1999], Feinstein [1998], Johnson

[1999], Jones and Tukey [2000], McBride, Loftis, and Adkins [1993], Nester [1996], Parkhurst [2001], and Suter [1996].

The vast majority of such cautions are unnecessary provided that we treat p values as merely one part of the evidence to be used in decision-making. They need to be viewed and interpreted in the light of all the surrounding evidence, past and present. No computer should be allowed to make decisions for you.

A failure to reject may result from insensitive or inappropriate measurements, or too small a sample size.

A difference that is statistically significant may be of no practical interest. Take a large enough sample and we will always reject the null hypothesis; take too small a sample and we will always accept—to say nothing of "significant" results that arise solely because their authors chose to test a "null" hypothesis rather than one of practical interest. (See Chapter 4.)

Many researchers would argue that there are always three regions to which a statistic may be assigned: acceptance, rejection, and indifference. When a statistic falls in the latter, intermediate region it may suggest a need for additional experiments. The p value is only one brick in the wall; all our other knowledge must and should be taken into consideration (Horwitz et al., 1998).

SUMMARY

- Provide details of power and sample size calculations.
- Describe treatment allocation.
- Detail exceptions including withdrawals and other sources of missing data.
- Use meaningful measures of dispersion.
- Use confidence intervals in preference to p values.
- Report sources of bias.
- Formal statistical inference is appropriate only for randomized studies and predetermined hypotheses.

TO LEARN MORE

The text by Lang and Secic [1997] is must reading; reporting criteria for meta-analyses are given on page 177 ff. See Tufte [1983] on the issue of table versus graph. For more on the geometric versus arithmetic mean see Parkhurst [1998]. For more on reporting requirements, see Begg et al. [1996], Bailar and Mosteller [1988], Grant [1989], Altman et al. [2001; the revised CONSORT statement], and International Committee of Medical Journal Editors [1997].

Mosteller [1979] and Anderson and Hauck [1986] warn against the failure to submit reports of negative or inconclusive studies and the failure of journal editors to accept them. To address this issue, the *Journal of Negative Results in Biomedicine* has just been launched at http://www.jnrbm.com/start.asp.

On the proper role of p values, see Neyman [1977], Cox [1977], http://www.coe.tamu.edu/~bthompson, http://www.indiana.edu/~stigsts, http://www.nprc.ucgs.gov/perm/hypotest, and Poole [1987, 2001].

To learn more about decision theory and regions of indifference, see Duggan and Dean [1968] and Hunter and Schmidt [1997].

Chapter 8
Graphics

KISS—Keep It Simple, but Scientific
Emanuel Parzen

What is the dimension of the information you will illustrate? Do you need to illustrate repeated information for several groups? Is a graphical illustration the best vehicle for communicating information to the reader? How do you select from a list of competing choices? How do you know whether the graphic you produce is effectively communicating the desired information?

GRAPHICS SHOULD EMPHASIZE AND HIGHLIGHT SALIENT FEATURES. THEY should reveal data properties and make large quantities of information coherent. While graphics provide the reader a break from dense prose, authors must not forget that their illustrations should be scientifically informative as well as decorative. In this chapter, we outline mistakes in selection, creation, and execution of graphics and discuss improvements for each of these three areas.

Graphical illustrations should be simple and pleasing to the eye, but the presentation must remain scientific. In other words, we want to avoid those graphical features that are purely decorative while keeping a critical eye open for opportunities to enhance the scientific inference we expect from the reader. A good graphical design should maximize the proportion of the ink used for communicating scientific information in the overall display.

THE SOCCER DATA

Dr. Hardin coaches youth soccer (players of age 5) and has collected the total number of goals for the top five teams during the eight-game spring 2001 season. The total number of goals scored per team was 16 (team 1), 22 (team 2), 14 (team 3), 11 (team 4), and 18 (team 5). There are many

Common Errors in Statistics (and How to Avoid Them), by Phillip I. Good and James W. Hardin.
ISBN 0-471-46068-0   Copyright © 2003 John Wiley & Sons, Inc.

ways we can describe this set of outcomes to the reader. In text above, we simply communicated the results in text.

A more effective presentation would be to write that the total number of goals scored by teams 1 through 5 was 16, 22, 14, 11, and 18, respectively. The College Station Soccer Club labeled the five teams as Team 1, Team 2, and so on. These labels show the remarkable lack of imagination that we encounter in many data collection efforts. Improving on this textual presentation, we could also say that the total number of goals with the team number identified by subscript was 22_2, 18_5, 16_1, 14_3, and 11_4. This presentation better communicates with the reader by ordering the outcomes because the reader will naturally want to know the order in this case.

FIVE RULES FOR AVOIDING BAD GRAPHICS

There are a number of choices in presenting the soccer outcomes in graphical form. Many of these are poor choices; they hide information, make it difficult to discern actual values, or inefficiently use the space within the graph. Open almost any newspaper and you will see a bar chart graphic similar to Figure 8.1 illustrating the soccer data. In this section,

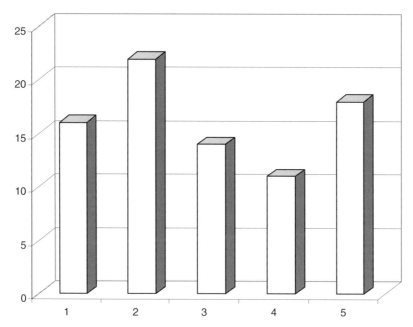

FIGURE 8.1 **Total Number of Goals Scored by Teams 1 through 5.** The x axis indicates the team number, and the y axis indicates the number of goals scored by the respective team. *Problem:* The false third dimension makes it difficult to discern values. The reader must focus on the top of the obscured back face to accurately interpret the values plotted.

we illustrate five important rules for generating correct graphics. Subsequent sections will augment this list with other specific examples.

Figure 8.1 includes a false third dimension; a depth dimension that does not correspond to any information in the data. Furthermore, the resulting figure makes it difficult to discern the actual values presented. Can you tell by looking at Figure 8.1 that Team 3 scored 14 goals, or does it appear that they scored 13 goals? The reader must focus on the top back corner of the three-dimensional rectangle since that part of the three-dimensional bar is (almost) at the same level as the grid lines on the plot; actually, the reader must first focus on the floor of the plot to initially discern the vertical distance of the back right corner of the rectangular bar from the corresponding grid line at the back (these are at the same height). The viewer must then mentally transfer this difference to the top of the rectangular bars in order to accurately infer the correct value. The reality is that most people focus on the front face of the rectangle and will subsequently misinterpret this data representation.

Figure 8.2 also includes a false third dimension. As before, the resulting illustration makes it difficult to discern the actual values presented. This illusion is further complicated by the fact that the depth dimension has been eliminated at the top of the three-dimensional pyramids so that it's nearly impossible to correctly ascertain the plotted values. Focus on the result of Team 4, compare it to the illustration in Figure 8.1, and judge whether you think the plots are using the same data (they are). Other types of plots that confuse the audience with false third dimensions include point plots with shadows and line plots where the data are connected with a three dimensional line or ribbon.

The lesson from these first two graphics is that we must avoid illustrations that utilize more dimensions than exist in the data. Clearly, a better presentation would indicate only two dimensions where one dimension identifies the teams and the other dimension identifies the number of goals scored.

> **Rule 1:** *Don't produce graphics illustrating more dimensions than exist in the data.*

Figure 8.3 is an improvement over three-dimensional displays. It is easier to discern the outcomes for the teams, but the axis label obscures the outcome of Team 4. Axes should be moved outside of the plotting area with enough labels so that the reader can quickly scan the illustration and identify values.

> **Rule 2:** *Don't superimpose labeling information on the graphical elements of interest. Labels can add information to the plot, but*

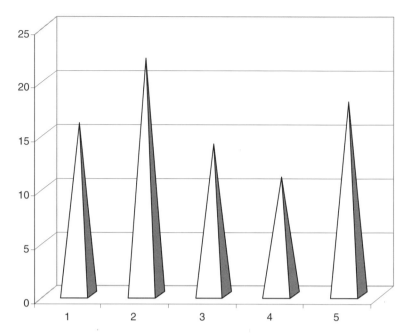

FIGURE 8.2 **Total Number of Goals Scored by Teams 1 through 5.** The *x* axis indicates the team number, and the *y* axis indicates the number of goals scored by the respective team. *Problem:* The false third dimension makes it difficult to discern the values in the plot. Since the back face is the most important for interpreting the values, the fact that the decorative object comes to a point makes it impossible to correctly read values from the plot.

> *should be placed in (otherwise) unused portions of the plotting region.*

Figure 8.4 is a much better display of the information of interest. The problem illustrated is that there is too much empty space in the graphic. Choosing to begin the vertical axis at zero means that about 40% of the plotting region is empty. Unless there is a scientific reason compelling you to include a specific baseline in the graph, the presentation should be limited to the range of the information at hand. There are several instances where axis range can exceed the information at hand, and we will illustrate those in a presentation.

> ***Rule 3:*** *Don't allow the range of the axes labels to significantly decrease the area devoted to data presentation. Choose axis limits wisely and do not automatically accept default values for the axes that are far outside of the range of data.*

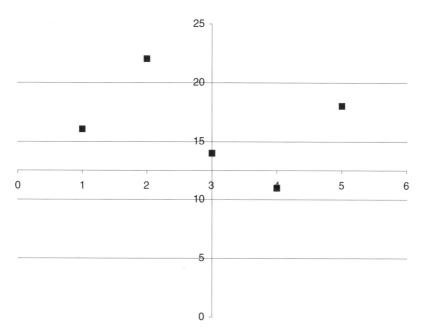

FIGURE 8.3 **Total Number of Goals Scored by Teams 1 through 5.** The *x* axis indicates the team number, and the *y* axis indicates the number of goals scored by the respective team. *Problem:* Placing the axes inside of the plotting area effectively occludes data information. This violates the simplicity goal of graphics; the reader should be able to easily see all of the numeric labels in the axes and plot region.

Figure 8.5 eliminates the extra space included in Figure 8.4 where the vertical axis is allowed to more closely match the range of the outcomes. The presentation is fine, but could be made better. The data of interest in this case involve a continuous and a categorical variable. This presentation treats the categorical variable as numeric for the purposes of organizing the display, but this is not necessary.

> *Rule 4: Carefully consider the nature of the information underlying the axes. Numeric axis labels imply a continuous range of values that can be confusing when the labels actually represent discrete values of an underlying categorical variable.*

Figures 8.5 and 8.6 are further improvements of the presentation. The graph region, area of the illustration devoted to the data, is illustrated with axes that more closely match the range of the data. Figure 8.6 connects the point information with a line that may help visualize the difference between the values, but also indicates a nonexistent relationship; the

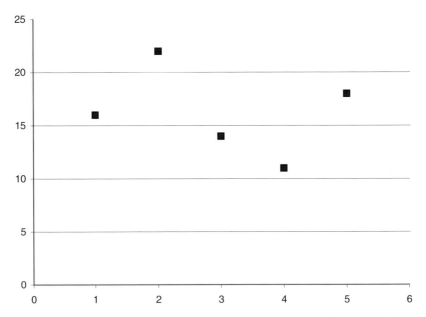

FIGURE 8.4 **Total Number of Goals Scored by Teams 1 through 5.** The x axis indicates the team number, and the y axis indicates the number of goals scored by the respective team. *Problem:* By allowing the y axis to range from zero, the presentation reduces the proportion of the plotting area in which we are interested. Less than half of the vertical area of the plotting region is used to communicate data.

horizontal axis is discrete rather than continuous. Even though these presentations vastly improve the illustration of the desired information, we are still using a two-dimensional presentation. In fact, our data are not really two-dimensional and the final illustration more accurately reflects the true nature of the information.

> **Rule 5:** *Do not connect discrete points unless there is either (a) a scientific meaning to the implied interpolation or (b) a collection of profiles for group level outcomes.*

Rules 4 and 5 are aimed at the practice of substituting numbers for labels and then treating those numeric labels as if they were in fact numeric. Had we included the word "Team" in front of the labels, there would be no confusion as to the nature of the labels. Even when nominative labels are used on an axis, we must consider the meaning of values between the labels. If the labels are truly discrete, data outcomes should not be connected or they may be misinterpreted as implying a continuous rather than discrete collection of values.

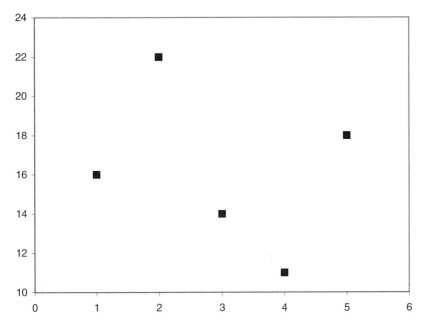

FIGURE 8.5 Total Number of Goals Scored by Teams 1 through 5. The *x* axis indicates the team number, and the *y* axis indicates the number of goals scored by the respective team. *Problem:* This graph correctly scales the *y* axis, but still uses a categorical variable denoting the team on the *x* axis. Labels 0 and 6 do not correspond to a team number and the presentation appears as if the *x* axis is a continuous range of values when in fact it is merely a collection of labels. While a reasonable approach to communicating the desired information, we can still improve on this presentation by changing the numeric labels on the *x* axis to String labels corresponding to the actual team names.

Figure 8.7 is the best illustration of the soccer data. There are no false dimensions, the range of the graphic is close to the range of the data, there is no difficulty interpreting the values indicated by the plotting symbols, and the legend fully explains the material. Alternatively, we can produce a simple table.

Table 8.1 succinctly presents the relevant information. Tables and figures have the advantage over in-text descriptions that the information is more easily found while scanning through the containing document. If the information is summary in nature, we should make that information easy to find for the reader and place it in a figure or table. If the information is ancillary to the discussion, it can be left in text.

Choosing Between Tabular and Graphical Presentations

In choosing between tabular and graphical presentations, there are two issues to consider: the size (density) of the resulting graphic and the scale

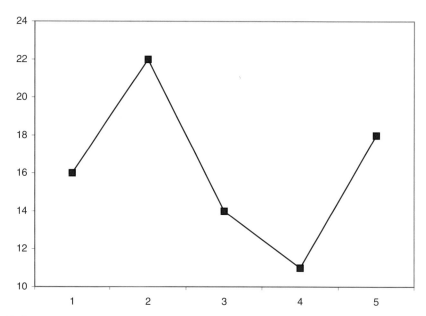

FIGURE 8.6 Total Number of Goals Scored by Teams 1 through 5. The x axis indicates the team number, and the y axis indicates the number of goals scored by the respective team. *Problem:* The inclusion of a polyline connecting the five outcomes helps the reader to visualize changes in scores. However, the categorical values are not ordinal, and the polyline indicates an interpolation of values that does not exist across the categorical variable denoting the team number. In other words, there is no reason that Team 5 is to the right of Team 3 other than we ordered them that way, and there is no Team 3.5 as the presentation seems to suggest.

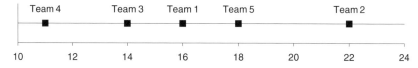

FIGURE 8.7 Total Number of Goals Scored by Teams 1 through 5. The x axis indicates with a square the number of goals scored by the respective team. The associated team name is indicated above the square. Labeling the outcomes addresses the science of the KISS specification given at the beginning of the chapter.

of the information. If the required number of rows for a tabular presentation would require more than one page, the graphical representation is preferred. Usually, if the amount of information is small, the table is preferred. If the scale of the information makes it difficult to discern otherwise significant differences, a graphical presentation is better.

TABLE 8.1 Total Number of Goals Scored by Teams 1 through 5 Ordered by Lowest Total to Highest Total[a]

Team 4	Team 3	Team 1	Team 5	Team 2
11	14	16	18	22

[a] These totals are for the Spring 2001 season. The organization of the table correctly sorts on the numeric variable. That the team labels are not sorted is far less important since these labels are merely nominal; were it not for the fact that we labeled with integers, the team names would have no natural ordering.

ONE RULE FOR CORRECT USAGE OF THREE-DIMENSIONAL GRAPHICS

As illustrated in the previous section, the introduction of superfluous dimensions in graphics should be avoided. The prevalence of turnkey solutions in software that implement these decorative presentations is alarming. At one time, these graphics were limited to business-oriented software and presentations, but this is no longer true. Misleading illustrations are starting to appear in scientific talks. This is partly due to the introduction of business-oriented software in university service courses (demanded by the served departments). Errors abound when increased license costs for scientific- and business-oriented software lead departments to eliminate the more scientifically oriented software packages.

The reader should not necessarily interpret these statements as a mandate to avoid business-oriented software. Many of these maligned packages are perfectly capable of producing scientific plots. Our warning is that we must educate ourselves in the correct software specifications.

Three-dimensional perspective plots are very effective, but require specification of a viewpoint. Experiment with various viewpoints to highlight the properties of interest. Mathematical functions lend themselves to three-dimensional plots, but raw data are typically better illustrated with contour plots. This is especially true for map data, such as surface temperatures, or surface wind (where arrows can denote direction and the length of the arrow can denote the strength).

In Figures 8.8 and 8.9, we illustrate population density of children for Harris County, Texas. Illustration of the data on a map is a natural approach, and a contour plot reveals the pockets of dense and sparse populations.

While the contour plot in Figure 8.8 lends itself to comparison of maps, the perspective plot in Figure 8.9 is more difficult to interpret. The surface is more clearly illustrated, but the surface itself prevents viewing all of the data.

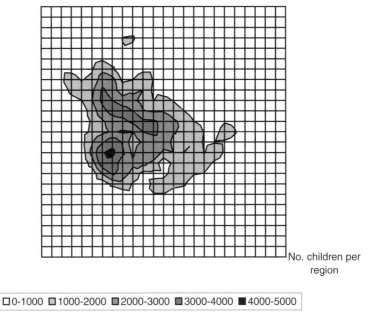

No. children per region

□ 0-1000 ▨ 1000-2000 ▩ 2000-3000 ▨ 3000-4000 ■ 4000-5000

FIGURE 8.8 **Distribution of Child Population in Harris County, Texas.**
The *x* axis is the longitude (−96.04 to −94.78 degrees), and the *y* axis is the
latitude (29.46 to 30.26 degrees).

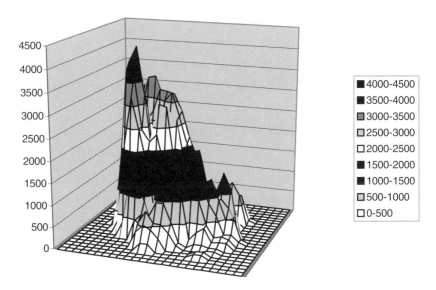

■ 4000-4500
■ 3500-4000
▨ 3000-3500
□ 2500-3000
□ 2000-2500
■ 1500-2000
■ 1000-1500
▨ 500-1000
□ 0-500

FIGURE 8.9 **Population Density of the Number of Children in Harris
County, Texas.** The *x* axis is the longitude (−96.04 to −94.78 degrees), and the
y axis is the latitude (29.46 to 30.26 degrees). The *x–y* axis is rotated 35 degrees
from Figure 8.10.

Rule 6: Use a contour plot over a perspective plot if a good viewpoint is not available. Always use a contour plot over the perspective plot when the axes denote map coordinates.

Though the contour plot is generally a better representation of mapped data, a desire to improve Figure 8.8 would lead us to suggest that the grid lines should be drawn in a lighter font so that they have less emphasis than lines for the data surface. Another improvement to data illustrated according to real-world maps is to overlay the contour plot where certain known places or geopolitical distinctions may be marked. The graphic designer must weigh the addition of such decorative items with the improvement in inference that they bring.

ONE RULE FOR THE MISUNDERSTOOD PIE CHART

The pie chart is undoubtedly the graphical illustration with the worst reputation. Wilkinson (1999) points out that the pie chart is simply a bar chart that has been converted to polar coordinates.

Focusing on Wilkinson's point makes it easier to understand that the conversion of the bar height to an angle on the pie chart is most effective when the bar height represents a proportion. If the bars do not have values where the sum of all bars is meaningful, the pie chart is a poor choice for presenting the information (cf. Figure 8.10).

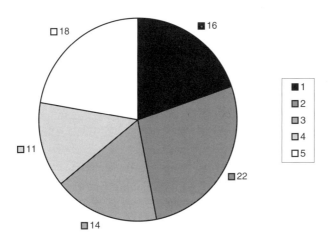

FIGURE 8.10 **Total Number of Goals Scored by Teams 1 through 5.** The legend indicates the team number and associated slice color for the number of goals scored by the respective team. The actual number of goals is also included. *Problem:* The sum of the individual values is not of interest so that the treatment of the individuals as proportions of a total is not correct.

Rule 7: Do not use pie charts unless the sum of the entries is scientifically meaningful and of interest to the reader.

On the other hand, the pie chart is an effective display for illustrating proportions. This is especially true when we want to focus on a particular slice of the graphic that is near 25% or 50% of the data since we humans are adept at judging these size portions. Including the actual value as a text element decorating the associated pie slice effectively allows us to communicate both the raw number along with the visual clue of the proportion of the total that the category represents. A pie chart intended to display information on all sections where some sections are very small is very difficult to interpret. In these cases, a table or bar chart is to be preferred.

Additional research has addressed whether the information should be ordered before placement in the pie chart display. There are no general rules to follow other than to repeat that humans are fairly good at identifying pie shapes that are one-half or one-quarter of the total display. As such, a good ordering of outcomes that included such values would strive to place the leading edge of 25% and 50% pie slices along one of the major north–south or east–west axes. Reordering the set of values may lead to confusion if all other illustrations of the data used a different ordering, so the graphic designer may ultimately feel compelled to reproduce other illustrations.

THREE RULES FOR EFFECTIVE DISPLAY OF SUBGROUP INFORMATION

Graphical displays are very effective for communication of subgroup information—for example, when we wish to compare changes in median family income over time of African-Americans and Hispanics. With a moderate number of subgroups, a graphical presentation can be much more effective than a similar tabular display. Labels, stacked bar displays, or a tabular arrangement of graphics can effectively display subgroup information. Each of these approaches has its limits, as we will see in the following sections.

In Figure 8.11, separate connected polylines easily separate the subgroup information. Each line is further distinguished with a different plotting symbol. Note how easy it is to confuse the information due to the inverted legend. To avoid this type of confusion, ensure that the order of entries (top to bottom) matches that of the graphic.

Rule 8: Put the legend items in the same order they appear in the graphic whenever possible.

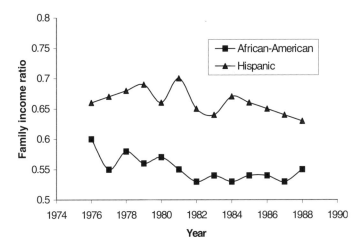

FIGURE 8.11 Median Family Income of African-Americans and Hispanics Divided by the Median Family Income for Anglo-American Families for Years 1976–1988. *Problem:* The legend identifies the two ethnic groups in the reverse order that they appear in the plot. It is easy to confuse the polylines due to the discrepancy in organizing the identifiers. The rule is that if the data follow a natural ordering in the plotting region, the legend should honor that order.

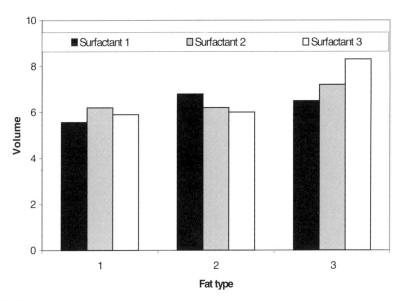

FIGURE 8.12 Volume of a Mixture Based on the Included fat and Surfactant Types. *Problem:* As with a scatterplot, the arbitrary decision to include zero on the *y* axis in a bar plot detracts from the focus on the values plotted.

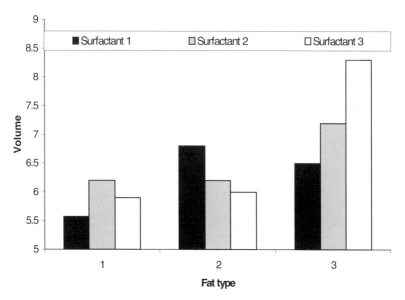

FIGURE 8.13 Volume of a Mixture Based on the Included fat and Surfactant Types. Drawing the bar plot with a more reasonable scale clearly distinguishes the values for the reader.

Clearly, there are other illustrations that would work even better for this particular data. When one subgroup is always greater than the other subgroup, we can use vertical bars between each measurement instead of two separate polylines. Such a display not only points out the discrepancies in the data, but also allows easier inference as to whether the discrepancy is static or changes over time.

The construction of a table such as Table 8.2 effectively reduces the number of dimensions from two to one. This presentation makes it more difficult for the reader to discern the subgroup information that the analysis emphasizes. While this organization matches the input to most statistical packages for correct analysis, it is not the best presentation for humans to discern the groups.

Keep in mind that tables are simply text-based graphics. All of the rules presented for graphical displays apply equally to textual displays.

The proper organization of the table in two dimensions clarifies the subgroup analysis. Tables may be augmented with decorative elements just as we augment graphics. Effective additions to the table are judged on their ability to focus attention on the science; otherwise these additions serve as distracters. Specific additions to tables include horizontal and vertical lines to differentiate subgroups, and font/color changes to distinguish headings from data entries.

TABLE 8.2 Volume of a Mixture Based on the Included Fat and Surfactant Types[a]

Fat	Surfactant	Volume
1	1	5.57
1	2	6.20
1	3	5.90
2	1	6.80
2	2	6.20
2	3	6.00
3	1	6.50
3	2	7.20
3	3	8.30

[a] **Problem:** The two categorical variables are equally of interest, but the table uses only one direction for displaying the values of the categories. This demonstrates that table generation is similar to graphics generation, and we should apply the same graphical rules honoring dimensions to tables.

TABLE 8.3 Volume of a Mixture Based on the Included Fat and Surfactant Types[a]

Fat	Surfactant		
	1	2	3
1	5.57	6.20	5.90
2	6.80	6.20	6.00
3	6.50	7.20	8.30

[a] The two categorical variables are equally of interest. With two categorical variables, the correct approach is to allow one to vary over rows and the other to vary over columns. This presentation is much better than the presentation of Table 8.2 and probably easier to interpret than any graphical representation.

Specifying a y axis that starts at zero obscures the differences of the results and violates Rule 3 seen previously. If we focus on the actual values of the subgroups, we can more readily see the differences.

TWO RULES FOR TEXT ELEMENTS IN GRAPHICS

If a picture were worth a thousand words, then the graphics we produce would considerably shorten our written reports. While attributing "a thousand words" for each graphic is an exaggeration, it remains true that the graphic is often much more efficient at communicating numeric information than equivalent prose. This efficiency is in terms of the amount of information successfully communicated and not necessarily any space savings.

If the graphic is a summary of numeric information, then the caption is a summary of the graphic. This textual element should be considered part of the graphic design and should be carefully constructed rather than placed as an afterthought. Readers, for their own use, often copy graphics and tables that appear in articles and reports. Failure on the part of the graphic designer to completely document the graphic in the caption can result in gross misrepresentation in these cases. It is not the presenter who copied the graph who suffers, but the original author who generated the graphic. Tufte [1983] advises that graphics "should be closely integrated with the statistical and verbal descriptions of the data set" and that the caption of the graphic clearly provides the best avenue for ensuring this integration.

> **Rule 9:** *Captions for your graphical presentations must be complete. Do not skimp on your descriptions.*

The most effective method for writing a caption is to show the graphic to a third party. Allow them to question the meaning and information presented. Finally, take your explanations and write them all down as a series of simple sentences for the caption. Readers rarely, if ever, complain that the caption is too long. If they do complain that the caption is too long, it is a clear indication that the graphic design is poor. Were the graphic more effective, the associated caption would be of a reasonable length.

Depending on the purpose of your report, editors may challenge the duplication of information within the caption and within the text. While we may not win every skirmish with those that want to abbreviate our reports, we are reminded that it is common for others to reproduce only tables and graphics from our reports for other purposes. Detailed captions help alleviate misrepresentations and other out-of-context references we certainly want to avoid, so we endeavor to win as many of these battles with editors as possible.

Other text elements that are important in graphical design are the axes labels, title, and symbols that can be replaced by textual identifiers. Recognizing that the plot region of the graph presents numerical data, the axis must declare associated units of measure. If the axis is transformed (log or otherwise), the associated label must present this information as well. The title should be short and serves as the title for the graphic and associated caption. By itself, the title usually does not contain enough information to fully interpret the graphic in isolation.

When symbols are used to denote points from the data that can be identified by meaningful labels, there are a few choices to consider for improving the information content of the graphic. First, we can replace all symbols with associated labels if such replacement results in a readable (nonoverlapping) presentation. If our focus highlights a few key points, we can substitute labels for only those values.

When replacing (or decorating) symbols with labels results in an overlapping indecipherable display, a legend is an effective tool provided that there are not too many legend entries. Producing a graphical legend with 100 entries is not an effective design. It is an easy task to design these elements when we stop to consider the purpose of the graphic. It is wise to consider two separate graphics when the amount of information overwhelms our ability to document elements in legends and the caption.

Too many line styles or plotting points can be visually confusing and prevent inference on the part of the reader. You are better off splitting the single graphic into multiple presentations when there are too many subgroups. An ad hoc rule of thumb is to limit the number of colors or symbols to less than eight.

Rule 10: Keep line styles, colors, and symbols to a minimum.

MULTIDIMENSIONAL DISPLAYS

Representing several distinct measures for a collection of points is problematic in both text and graphics. The construction of tables for this display is difficult due to the necessity of effectively communicating the array of subtabular information. The same is true in graphical displays, but the distinction of the various quantities is somewhat easier.

CHOOSING EFFECTIVE DISPLAY ELEMENTS

As Cleveland and McGill (1988) emphasize, graphics involve both encoding of information by the graphic designer and decoding of the information by the reader. Various psychological properties affect the decoding of the information in terms of the reader's graphical perception. For example, when two or more elements are presented, the reader will also envision byproducts such as implied texture and shading. These byproducts can be distracting and even misleading.

Graphical displays represent a choice on the part of the designer in terms of the quantitative information that is highlighted. These decisions are based on the desire to assist the analyst and reader in discerning performance and properties of the data and associated models fitted to the data. While many of the decisions in graphical construction simply follow convention, the designer is still free to choose geometric shapes to represent points, color or style for lines, and shading or textures to represent areas. The referenced authors included a helpful study in which various graphical styles were presented to readers. The ability to discern the underlying infor-

TABLE 8.4 Rank-Ordered List of Elementary Design Choices for Conveying Numeric Information[a]

Rank	Graphical Element[b]
1	Positions along a common scale
2	Positions along identical, nonaligned scales
3	Lengths
4	Angles
4–10	Slopes
6	Areas
7	Volumes
8	Densities
9	Color saturations
10	Color hues

[a] Slopes are given a wide range of ranks since they can be very poor choices when the aspect ratio of the plot does not allow distinction of slopes. Areas and volumes introduce false dimensions to the display that prevent readers from effective interpretation of the underlying information.
[b] Graphical elements are ordered from most (1) to least (10) effective.

mation was measured for each style, and an ordered list of effective elementary design choices was inferred. The ordered list for illustrating numeric information is presented in Table 8.4. The goal of the list is to allow the reader to effectively differentiate among several values.

CHOOSING GRAPHICAL DISPLAYS

When relying completely on the ability of software to produce scientific displays, many authors are limited by their mastery of the software. Most software packages will allow users to either (a) specify in advance the desired properties of the graph or (b) edit the graph to change individual items in the graph. Our ability to follow the guidelines outlined in this chapter is directly related to the time we spend learning to use the more advanced graphics features of software.

SUMMARY

- Examine the data and results to determine the number of dimensions in the information to be illustrated. Limit your graphic to that many dimensions.
- Limit the axes to exactly (or closely) match the range of data in the presentation.
- Do not connect points in a scatterplot unless there is an underlying interpolation that makes scientific sense.

- Recognize that readers of your reports will copy tables and figures for their own use. Ensure that you are not misquoted by completely describing your graphics and tables in the associated legends. Do not skimp on these descriptions or you force readers to scan the entire document for needed explanations.

- If readers are to accurately compare two different graphics for values (instead of shapes or predominant placement of outcomes), use the same axis ranges on the two plots.

- Use pie charts only when there are a small number of categories and the sum of the categorical values has scientific meaning.

- Tables are text-based graphics. Therefore, the rules governing organization and scientific presentation of graphics should be honored for the tables that we present. Headings should be differentiated from data entries by font weight or color change. Refrain from introducing multiple fonts in the tables and instead use one font where differences are denoted in weight (boldness), style (slanted), and size.

- Numeric entries in tables should be in the same number of significant digits. Furthermore, they should be right justified so that they line up and allow easy interpretation while scanning columns of numbers.

- Many of the charts could benefit from the addition of grid lines. Bar charts especially can benefit from horizontal grid lines from the y-axis labels. This is especially true of wider displays, but grid lines should be drawn in a lighter shade than the lines used to draw the major features of the graphic.

- Criticize your graphics and tables after production by isolating them with their associated caption. Determine if the salient information is obvious by asking a colleague to interpret the display. If we are serious about producing efficient communicative graphics, we must take the time ensure that our graphics are interpretable.

TO LEARN MORE

Wilkinson (1999) presents a formal grammar for describing graphics, but more importantly (for our purposes), the author lists graphical element hierarchies from best to worst. Cleveland (1985) focuses on the elements of common illustrations where he explores the effectiveness of each element in communicating numeric information. A classic text is Tukey (1977), where the author lists both graphical and text-based graphical summaries of data. More recently, Tufte (1983, 1990) organized much of the previous work and combined that work with modern developments. For specific illustrations, subject-specific texts can be consulted for particular displays in context; for example, Hardin and Hilbe (2003, pp. 143–167) illustrate the use of graphics for assessing model accuracy.

Part III
BUILDING A MODEL

Chapter 9

Univariate Regression

Are the data adequate? Does your data set cover the entire range of interest? Will your model depend on one or two isolated data points?

THE SIMPLEST EXAMPLE OF A MODEL, THE RELATIONSHIP between exactly two variables, illustrates at least five of the many complications that can interfere with the task of model building:

1. Limited scope—the model we develop may be applicable for only a portion of the range of each variable.

2. Ambiguous form of the relationship—a variable may give rise to a statistically significant linear regression without the underlying relationship being a straight line.

3. Confounding—undefined confounding variables may create the illusion of a relationship or may mask an existing one.

4. Assumptions—the assumptions underlying the statistical procedures we use may not be satisfied.

5. Inadequacy—goodness of fit is not the same as prediction.

We consider each of these error sources in turn along with a series of preventive measures. Our discussion is divided into problems connected with model selection and difficulties that arise during the estimation of model coefficients.

MODEL SELECTION

Limited Scope

Almost every relationship has both a linear and a nonlinear portion where the nonlinear portion is increasingly evident for both extremely large and

Common Errors in Statistics (and How to Avoid Them), by Phillip I. Good and James W. Hardin. ISBN 0-471-46068-0 Copyright © 2003 John Wiley & Sons, Inc.

extremely small values. One can think of many examples from physics such as Boyle's Law, which fails at high pressures, and particle symmetries that are broken as the temperature falls. In medicine, radio immune assay fails to deliver reliable readings at very low dilutions and for virtually every drug there will always be an increasing portion of nonresponders as the dosage drops. In fact, almost every measuring device—electrical, electronic, mechanical, or biological—is reliable only in the central portion of its scale.

We need to recognize that while a regression equation may be used for interpolation within the range of measured values, we are on shaky ground if we try to extrapolate, to make predictions for conditions not previously investigated. The solution is to know the range of application and to recognize, even if we do not exactly know the range, that our equations will be applicable to some but not all possibilities.

Ambiguous Relationships

Think why rather than what.

The exact nature of the formula connecting two variables cannot be determined by statistical methods alone. If a linear relationship exists between two variables X and Y, then a linear relationship also exists between Y and any monotone (nondecreasing or nonincreasing) function of X. Assume that X can only take positive values. If we can fit Model I: $Y = \alpha + \beta X + \varepsilon$ to the data, we also can fit Model II: $Y = \alpha' + \beta' \log[X] + \varepsilon$, and Model III: $Y = \alpha'' + \beta'' X + \gamma X^2 + \varepsilon$. It can be very difficult to determine which model, if any, is the "correct" one in either a predictive or mechanistic sense.

A graph of Model I is a straight line (see Figure 9.1). Because Y includes a stochastic or random component ε, the pairs of observations (x_1, y_1), (x_2, y_2), ... will not fall exactly on this line but above and below it. The function $\log[X]$ does not increase as rapidly as X does; when we fit Model II to these same pairs of observations, its graph rises above that of Model I for small values of X and falls below that of Model I for large values. Depending on the set of observations, Model II may give just as good a fit to the data as Model I.

How Model III behaves will depend upon whether β'' and α'' are both positive or whether one is positive and the other negative. If β'' and α'' are both positive, then the graph of Model III will lie below the graph of Model I for small positive values of X and above it for large values. If β'' is positive and α'' is negative, then Model III will behave more like Model II. Thus Model III is more flexible than either Models I or II and can usually be made to give a better fit to the data—that is, to minimize some

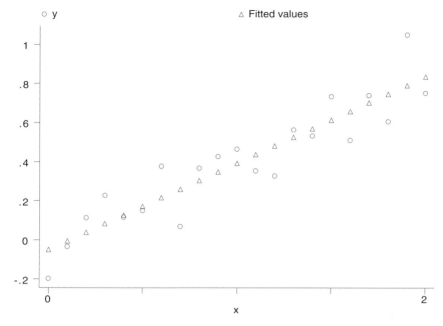

FIGURE 9.1 A Straight Line Appears to Fit the Data.

function of the differences between what is observed, y_i, and what is predicted by the model, $Y[x_i]$.

The coefficients α, β, γ for all three models can be estimated by a technique known (to statisticians) as linear regression. Our knowledge of this technique should not blind us to the possibility that the true underlying model may require nonlinear estimation as in

$$\text{Model IV:} \quad Y = \frac{\alpha + \beta X + \gamma X^2}{\delta - \phi X} + \varepsilon.$$

This latter model may have the advantage over the first three in that it fits the data over a wider range of values.

Which model should we choose? At least two contradictory rules apply:

- The more parameters, the better the fit; thus, Model III and Model IV are to be preferred.
- The simpler, more straightforward model is more likely to be correct when we come to apply it to data other than the observations in hand; thus, Models I and II are to be preferred.

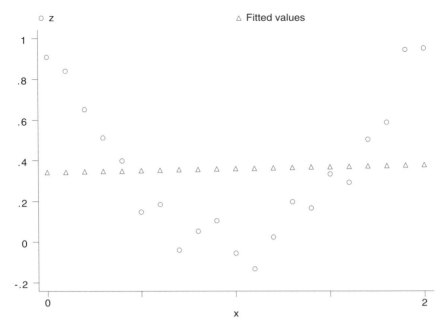

FIGURE 9.2 Fitting an Inappropriate Model.

Again, the best rule of all is not to let statistics do your thinking for you, but to inquire into the mechanisms that give rise to the data and that might account for the relationship between the variables X and Y. An example taken from physics is the relationship between volume V and temperature T of a gas. All of the preceding four models could be used to fit the relationship. But only one, the model $V = a + KT$, is consistent with Kinetic Molecular Theory.

Inappropriate Models

An example in which the simpler, more straightforward model is not correct comes when we try to fit a straight line to what is actually a higher-order polynomial. For example, suppose we tried to fit a straight line to the relationship $Y = (X - 1)^2$ over the range $X = (0,+2)$. We'd get a line with slope 0 similar to that depicted in Figure 9.2. With a correlation of 0, we might even conclude in error that X and Y were not related. Figure 9.2 suggests a way we can avoid falling into a similar trap. **Always plot the data before deciding on a model.**

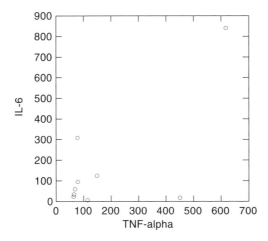

FIGURE 9.3 Relation Between Two Inflammatory Reaction Mediators in Response to Silicone Exposure. Data taken from Mena et al. [1995].

The data in Figure 9.3 are taken from Mena et al. [1995]. These authors reported in their abstract that, "The correlation . . . between IL-6 and TNF-alpha was .77 . . . statistically significant at a *p*-value less than .01." Would you have reached the same conclusion?

With more complicated models, particularly those like Model IV that are nonlinear, it is advisable to calculate several values that fall outside the observed range. If the results appear to defy common sense (or the laws of physics, market forces, etc.), the nonlinear approach should be abandoned and a simpler model utilized.

Often it can be difficult to distinguish which variable is the cause and which one is the effect. But if the values of one of the variables are fixed in advance, then this variable should always be treated as the so-called independent variable or cause, the *X* in the equation $Y = a + bX + \varepsilon$. Here is why:

When we write $Y = a + bx + \varepsilon$, we actually mean $Y = E(Y|x) + \varepsilon$, where $E(Y|X) = a + bx$ is the expected value of an indefinite number of independent observations of *Y* when $X = x$. If *X* is fixed, the inverse equation $x = (E(x|Y) - a)/b + \varepsilon' =$ makes little sense.

Confounding Variables

If the effects of additional variables other than *X* on *Y* are suspected, these additional effects should be accounted for either by stratifying or by performing a multivariate regression.

SOLVE THE RIGHT PROBLEM

Don't be too quick to turn on the computer. Bypassing the brain to compute by reflex is a sure recipe for disaster.

Be sure of your objectives for the model: Are you trying to uncover cause-and-effect mechanisms? Or derive a formula for use in predictions? If the former is your objective, standard regression methods may not be appropriate.

A researcher studying how neighborhood poverty levels affect violent crime rates hit an apparent statistical roadblock. Some important criminological theories suggest that this positive relationship is curvilinear with an accelerating slope while other theories suggest a decelerating slope. As the crime data are highly variable, previous analyses had used the logarithm of the primary end point—violent crime rate—and reported a significant negative quadratic term (poverty*poverty) in their least-squares models. The researcher felt that such results were suspect, that the log transformation alone might have biased the results toward finding a significant negative quadratic term for poverty.

But quadratic terms and log transforms are irrelevancies, artifacts result-ing from an attempt to squeeze the data into the confines of a linear regression model. The issue appears to be whether the rate of change of crime rates with poverty levels is a constant, increasing, or decreasing function of poverty levels. Resolution of this issue requires a totally differ-ent approach.

Suppose Y denotes the variable you are trying to predict and X denotes the predictor. Replace each of the $y[i]$ by the slope $y^*[i] = (y[i + 1] - y[i])/(x[i + 1] - x[i])$. Replace each of the $x[i]$ by the midpoint of the inter-val over which the slope is measured, $x^*[i] = (x[i + 1] - x[i])/2$. Use the permutation methods described in Chapter 5 to test for the correlation if any between y^* and x^*. A positive correlation means an accelerating slope, a negative correlation, a decelerating slope.

Correlations can be deceptive. Variable X can have a statistically significant correlation with variable Y, solely because X and Y are both dependent on a third variable Z. A fall in the price of corn is inversely proportional to the number of hay-fever cases only because the weather that produces a bumper crop of corn generally yields a bumper crop of ragweed as well.

Even if the causal force X under consideration has no influence on the dependent variable Y, the effects of unmeasured selective processes can produce an apparent test effect. Children were once taught that storks brought babies. This juxtaposition of bird and baby makes sense (at least to a child) because where there are houses there are both families and chimneys where storks can nest. The bad air or miasma model ("common sense" two centuries ago) works rather well at explaining respiratory ill-nesses and not at all at explaining intestinal ones. An understanding of the role that bacteria and viruses play unites the two types of illness and enriches our understanding of both.

We often try to turn such pseudo-correlations to advantage in our research, using readily measured *proxy variables* in place of their less easily measured "causes." Examples are our use of population change in place of economic growth, M2 for the desire to invest, arm cuff blood pressure measurement in place of the width of the arterial lumen, and tumor size for mortality. At best, such *surrogate responses* are inadequate (as in attempting to predict changes in stock prices); in other instances they may actually point in the wrong direction.

At one time, the level of CD-4 lymphocytes in the blood appeared to be associated with the severity of AIDs; the result was that a number of clinical trials used changes in this level as an indicator of disease status. Reviewing the results of 16 sets of such trials, Fleming [1995] found that the concentration of CD-4 rose to favorable levels in 13 instances even though clinical outcomes were only favorable in eight.

Stratification

Gender discrimination lawsuits based on the discrepancy in pay between men and women could be defeated once it was realized that pay was related to years in service and that women who had only recently arrived on the job market in great numbers simply didn't have as many years on the job as men.

These same discrimination lawsuits could be won once the gender comparison was made on a years-in-service basis—that is, when the salaries of new female employees were compared with those of newly employed men, when the salaries of women with three years of service were compared with those of men with the same time in grade, and so forth. Within each stratum, men always had the higher salaries.

If the effects of additional variables other than X on Y are suspected, they should be accounted for either by stratifying or by performing a multivariate regression as described in the next chapter.

The two approaches are *not* equivalent unless *all* terms are included in the multivariate model. Suppose we want to account for the possible effects of gender. Let $I[\ ]$ be an indicator function that takes the value 1 if its argument is true and 0 otherwise. Then to duplicate the effects of stratification, we would have to write the multivariate model in the following form:

$$Y = a_m I[\text{male}] + a_f (1 - I[\text{male}]) + b_m I[\text{male}]X + b_f (1 - I[\text{male}]) + e.$$

In a study by Kanarek et al. [1980], whose primary focus is the relation between asbestos in drinking water and cancer, results are stratified by sex, race, and census tract. Regression is used to adjust for income, education, marital status, and occupational exposure.

Lieberson [1985] warns that if the strata differ in the levels of some third unmeasured factor that influences the outcome variable, the results may be bogus.

Simpson's Paradox

A third omitted variable may also result in two variables appearing to be independent when the opposite is true. Consider the following table, an example of what is termed Simpson's paradox:

	Treatment Group	
	Control	Treated
Alive	6	20
Dead	6	20

We don't need a computer program to tell us the treatment has no effect on the death rate. Or does it? Consider the following two tables that result when we examine the males and females separately:

	Treatment Group	
	Control	Treated
Alive	4	8
Dead	3	5

	Treatment Group	
	Control	Treated
Alive	2	12
Dead	3	15

In the first of these tables, treatment reduces the male death rate from 3 out of 7 (0.43) to 5 out of 13 (0.38). In the second, the rate is reduced from 3 out of 5 (0.6) to 15 out of 27 (0.55). Both sexes show a reduction, yet the combined population does not. Resolution of this paradox is accomplished by avoiding a knee-jerk response to statistical significance when association is involved. One needs to think deeply about underlying cause-and-effect relationships before analyzing data. Thinking about cause and effect in the preceding example might have led us to think about possible sexual differences and to stratify the data by sex before analyzing it.

ESTIMATING COEFFICIENTS

Write down and confirm your assumptions before you begin.

In this section we consider problems and solutions associated with three related challenges:

1. **Estimating the coefficients of a model.**
2. **Testing hypotheses concerning the coefficients.**
3. **Estimating the precision of our estimates.**

The techniques we employ will depend upon the following:

1. **The nature of the regression function (linear, nonlinear, logistic).**
2. **The nature of the losses associated with applying the model.**
3. **The distribution of the error terms in the model—that is, the ε's.**
4. **Whether these error terms are independent or dependent.**

The estimates we obtain will depend upon our choice of fitting function. Our choice should not be dictated by the software but by the nature of the losses associated with applying the model. Our software may specify a least-squares fit—most commercially available statistical packages do—but our real concern may be with minimizing the sum of the absolute values of the prediction errors or the maximum loss to which one will be exposed.

Algorithms for least absolute deviation (LAD) regression are given in Barrodale and Roberts [1973]. The **qreg** function of Stata provides for LAD regression. The Blossom package available as freeware from http://www.mesc.usgs.gov/blossom/blossom.html includes procedures for LAD and quantile regression.

In the *univariate* linear regression model, we assume that

$$y = E(Y\,|\,x) + \varepsilon$$

where E denotes the mathematical expectation of Y given x and could be any deterministic function of x in which the parameters appear in linear form. ε, the error term, stands for all the other unaccounted for factors that make up the observed value y.

How accurate our estimates are and how consistent they will be from sample to sample will depend upon the nature of the error terms. If none of the many factors that contribute to the value of ε make more than a small contribution to the total, then ε will have a Gaussian distribution. If the $\{\varepsilon_i\}$ are independent and normally distributed (Gaussian), then the ordinary least-squares estimates of the coefficients produced by most statistical software will be unbiased and have minimum variance.

These desirable properties, indeed the ability to obtain coefficient values that are of use in practical applications, will not be present if the wrong model has been adopted. They will not be present if successive observations are dependent. The values of the coefficients produced by the software will not be of use if the associated losses depend on some function of the observations other than the sum of the squares of the differences between what is observed and what is predicted. In many practical problems, one is more concerned with minimizing the sum of the absolute values of the differences or with minimizing the maximum prediction error. Finally, if the error terms come from a distribution that is far from Gaussian, a distribution that is truncated, flattened, or asymmetric, the p values and precision estimates produced by the software may be far from correct.

Alternatively, we may use permutation methods to test for the significance of the resulting coefficients. Provided that the $\{\varepsilon_i\}$ are independent and identically distributed (Gaussian or not), the resulting p values will be exact. They will be exact regardless of which goodness-of-fit criterion is employed.

Suppose that our hypothesis is that $y_i = a + bx_i + \varepsilon_i$ for all i and $b = b_0$. First, we substitute $y'_1 = y_i - b_0 x_i$ in place of the original observations y_i. Our translated hypothesis is $y'_i = a + b'x_i + \varepsilon_i$ for all i and $b' = 0$ or, equivalently, $\rho = 0$, where ρ is the correlation between the variables Y' and X. Our test for correlation is based on the permutation distribution of the sum of the cross-products $y'_i x_i$ (Pitman, 1938). Alternative tests based on permutations include those of Cade and Richards [1996], and tests based on MRPP LAD regression include those of Mielke and Berry [1997].

For large samples, these tests are every bit as sensitive as the least-squares test described in the previous paragraph even when all the conditions for applying that test are satisfied (Mielke and Berry, 2001, Section 5.4).

If the errors are dependent and normally distributed and the covariances are the same for every pair of errors, then we may also apply any of the permutation methods described above. If the errors are dependent and normally distributed, but we are reluctant to make such a strong assumption about the covariances, then our analysis may call for dynamic regression models (Pankratz, 1991).[1]

FURTHER CONSIDERATIONS

Bad Data

The presence of bad data can completely distort regression calculations. When least-squares methods are employed, a single outlier can influence

---

[1] In the SAS manual, these are called ARIMAX techniques and are incorporated in Proc ARIMA.

the entire line to pass closely to the outlier. While a number of methods exist for detecting the most influential observations (see, for example, Mosteller and Tukey, 1977), influential does not automatically mean that the data point is in error. Measures of influence encourage review of data for exclusion. Statistics do not exclude data, analysts do. And they only exclude data when presented firm evidence that the data are in error.

The problem of bad data is particularly acute in two instances:

1. **When most of the data are at one end of the line, so that a few observations at the far end can have undue influence on the estimated model.**
2. **When there is no causal relationship between X and Y.**

The Washington State Department of Social and Health Services extrapolates its audit results on the basis of a regression of over- and undercharges against the dollar amount of the claim. Because the frequency of errors depends on the amount of paper work involved and not on the dollar amount of the claim, no linear relationship exists between overcharges and the amount of the claim. The slope of the regression line can vary widely from sample to sample; the removal or addition of a very few samples to the original audit can dramatically affect the amount claimed by the state in overcharges.

Recommended is the *delete-one* approach in which the regression coefficients are recomputed repeatedly deleting a single pair of observations from the original data set each time. These calculations provide confidence intervals for the estimates along with an estimate of the sensitivity of the regression to outliers. When the number of data pairs exceeds 100, a bootstrap might be used instead.

To get an estimate of the precision of the estimates and the sensitivity of the regression equation to bad data, recompute the coefficients leaving out a different data pair each time.

Convenience

More often than we would like to admit, the variables and data that go into our models are chosen for us. We cannot directly measure the variables we are interested in, so we make do with surrogates. But such surrogates may or may not be directly related to the variables of interest. Lack of funds and/or the necessary instrumentation limit the range over which observations can be made. Our census overlooks the homeless, the uncooperative, and the less luminous. (See, for example, *City of New York v. Dept of Commerce*,[2] Disney [1976], and Bothun [1998, Chapter 6].)

---

[2] 822 F. Supp. 906 (E.D.N.Y., 1993).

The presence of such bias does not mean we should abandon our attempts at modeling, but that we should be aware of and report our limitations.

Stationarity

An underlying assumption of regression methods is that relationships among variables remain constant during the data collection period. If not, if the variables we are measuring undergo seasonal or other detectable changes, then we need to account for them. A multivariate approach is called for as described in the next chapter.

Practical Versus Statistical Significance

An association can be of statistical significance without being of the least practical value. In the study by Kanarek et al. [1980] referenced above, a 100-fold increase in asbestos fiber concentration is associated with perhaps a 5% increase in lung cancer rates. Do we care? Perhaps, because no life can be considered unimportant. But courts traditionally have looked for at least a twofold increase in incidence before awarding damages. (See, for example, the citations in Chapter 6 of Good, 2001b.) And in this particular study, there is reason to believe there might be other hidden cofactors that are at least as important as the presence of asbestos fiber.

Goodness-of-Fit Versus Prediction

As noted above, we have a choice of "fitting methods." We can minimize the sum of the squares of the deviations between the observed and model values, or we can minimize the sum of the absolute values of these deviations, or we can minimize some entirely different function. Suppose that we have followed the advice given above and have chosen our goodness-of-fit criterion to be identical with our loss function.

For example, suppose the losses are proportional to the square of the prediction errors, and we have chosen our model's parameters so as to minimize the sum of squares of the differences $y_i - M[x_i]$ for the historical data. Unfortunately, minimizing this sum of squares is no guarantee that when we continue to make observations, we will continue to minimize the sum of squares between what we observe and what our model predicts. If you are a businessman whose objective is to predict market response, this distinction can be critical.

There are at least three reasons for the possible disparity:

1. The original correlation was spurious.
2. The original correlation was genuine but the sample was not representative.
3. The original correlation was genuine, but the nature of the relationship has changed with time (as a result of changes in the

underlying politic, market, or environment, for example). We take up this problem again in our chapter on prediction error.

And lest we forget: Association does not "prove" causation, it can only contribute to the evidence.

Indicator Variables
The use of an indicator (yes/no) or a nonmetric ordinal variable (improved, much improved, no change) as the sole independent (X) variable is inappropriate. The two-sample and k-sample procedures described in Chapter 5 should be employed.

Transformations
It is often the case that the magnitude of the residual error is proportional to the size of the observations; that is, $y = E(Y|x)\varepsilon$. A preliminary log transformation will restore the problem to linear form $\log(y) = \log E(Y|x) + \varepsilon'$. Unfortunately, even if ε is normal, ε' is not, and the resulting confidence intervals need to be adjusted (Zhou and Gao, 1997).

Curve-Fitting and Magic Beans
Until recently, what distinguished statistics from the other branches of mathematics was that at least one aspect of each analysis was firmly grounded in reality. Samples were drawn from real populations and, in theory, one could assess and validate findings by examining larger and larger samples taken from that same population.

In this reality-based context, modeling has one or possibly both of the following objectives:

1. To better understand the mechanisms leading to particular responses.
2. To predict future outcomes.

Failure to achieve these objectives has measurable losses. While these losses cannot be eliminated because of the variation inherent in the underlying processes, it is hoped that by use of the appropriate statistical procedure, they can be minimized.

By contrast, the goals of curve fitting (nonparametric or local regression)[3] are aesthetic in nature; the resultant graphs, though pleasing to the eye, may bear little relation to the processes under investigation. To quote Green and Silverman [1994, p. 50], "there are two aims in curve estimation, which to some extent conflict with one another, to maximize goodness-of-fit and to minimize roughness."

---

[3] See, for example Green and Silverman [1994] and Loader [1999].

The first of these aims is appropriate *if* the loss function is mean-square error.[4] The second creates a strong risk of overfitting. Validation is essential, yet most of the methods discussed in Chapter 11 do not apply. Validation via a completely independent data set cannot provide confirmation, because the new data would entail the production of a completely different, unrelated curve. The only effective method of validation is to divide the data set in half at random, fit a curve to one of the halves, and then assess its fit against the entire data set.

SUMMARY

Regression methods work well with physical models. The relevant variables are known and so are the functional forms of the equations connecting them. Measurement can be done to high precision, and much is known about the nature of the errors—in the measurements and in the equations. Furthermore, there is ample opportunity for comparing predictions to reality.

Regression methods can be less successful for biological and social science applications. Before undertaking a univariate regression, you should have a fairly clear idea of the mechanistic nature of the relationship (and thus the form the regression function will take). Look for deviations from the model particularly at the extremes of the variable range. A plot of the residuals can be helpful in this regard; see, for example, Davison and Snell [1991] and Hardin and Hilbe [2003, pp. 143–159].

A preliminary multivariate analysis (the topic of the next two chapters) will give you a fairly clear notion of which variables are likely to be confounded so that you can correct for them by stratification. Stratification will also allow you to take advantage of permutation methods that are to be preferred in instances where "errors" or model residuals are unlikely to follow a normal distribution.

It's also essential that you have firmly in mind the objectives of your analysis, and the losses associated with potential decisions, so that you can adopt the appropriate method of goodness of fit. The results of a regression analysis should be treated with care; as Freedman [1999] notes, "Even if significance can be determined and the null hypothesis rejected or accepted, there is a much deeper problem. To make causal inferences, it must in essence be assumed that equations are invariant under proposed interventions. . . . if the coefficients and error terms change when the variables on the right hand side of the equation are manipulated rather than being passively observed, then the equation has only a limited utility for predicting the results of interventions."

---

[4] Most published methods also require that the residuals be normally distributed.

Statistically significant findings should serve as a motivation for further corroborative and collateral research rather than as a basis for conclusions.

Checklist: Write down and confirm your assumptions before you begin.

- Data cover an adequate range. Slope of line not dependent on a few isolated values.
- Model is plausible and has or suggests a causal basis.
- Relationships among variables remained unchanged during the data collection period and will remain unchanged in the near future.
- Uncontrolled variables are accounted for.
- Loss function is known and will be used to determine the goodness of fit criteria.
- Observations are independent, or the form of the dependence is known or is a focus of the investigation.
- Regression method is appropriate for the types of data involved and the nature of the relationship.
- Is the distribution of residual errors known?

TO LEARN MORE

David Freedman's [1999] article on association and causation is must reading. Lieberson [1985] has many examples of spurious association. Friedman, Furberg and DeMets [1996] cite a number of examples of clinical trials using misleading surrogate variables.

Mosteller and Tukey [1977] expand on many of the points raised here concerning the limitations of linear regression. Mielke and Berry [2001, Section 5.4] provide a comparison of MRPP, Cade-Richards, and OLS regression methods. Distribution-free methods for comparing regression lines among strata are described by Good [2001, pp. 168–169].

For more on Simpson's paradox, see
http://www.cawtech.freeserve.co.uk/simpsons.2.html. For a real-world example, search under Simpson's paradox for an analysis of racial bias in New Zealand Jury Service at http://www.stats.govt.nz.

Chapter 10
Multivariable Regression

MULTIVARIABLE REGRESSION IS PLAGUED BY THE SAME PROBLEMS univariate regression is heir to, plus many more of its own. Is the model correct? Are the associations spurious?

In the univariate case, if the errors were not normally distributed, we could take advantage of permutation methods to obtain exact significance levels in tests of the coefficients. Exact permutation methods do not exist in the multivariable case.

When selecting variables to incorporate in a multivariable model, we are forced to perform repeated tests of hypotheses, so that the resultant p values are no longer meaningful. One solution, if sufficient data are available, is to divide the data set into two parts, using the first part to select variables and using the second part to test these same variables for significance.

If choosing the correct functional form of a model in a univariate case presents difficulties, consider that in the case of k variables, there are k linear terms (should we use logarithms? should we add polynomial terms?) and $k(k - 1)$ first-order cross products of the form $x_i x_k$. Should we include any of the $k(k - 1)(k - 2)$ second-order cross products?

Should we use forward stepwise regression, or backward, or some other method for selecting variables for inclusion? The order of selection can result in major differences in the final form of the model (see, for example, Roy [1958] and Goldberger [1961]).

David Freedman [1983] searched for and found a large and highly significant R^2 among *totally independent* normally distributed random variables. This article is reproduced in its entirety in Appendix A, and we urge you to read this material more than once. Freedman demonstrates how

Common Errors in Statistics (and How to Avoid Them), by Phillip I. Good and James W. Hardin.
ISBN 0-471-46068-0   Copyright © 2003 John Wiley & Sons, Inc.

the testing of multiple hypotheses, a process that typifies the method of stepwise regression, can only exacerbate the effects of spurious correlation. As he notes in the introduction to the article, "If the number of variables is comparable to the number of data points, and if the variables are only imperfectly correlated among themselves, then a very modest search procedure will produce an equation with a relatively small number of explanatory variables, most of which come in with significant coefficients, and a highly significant R^2. This will be so even if Y is totally unrelated to the X's"

Freedman used computer simulation to generate 5100 independent normally distributed "observations." He put these values into a data matrix in the form required by the SAS regression procedure. His organization of the values defined 100 "observations" on each of 51 random variables. Arbitrarily, the first 50 variables were designated as "explanatory" and the 51st as the dependent variable Y.

In the first of two passes through the "data," all 50 of the explanatory variables were used. 15 coefficients out of the 50 were significant at the 25% level, and one out of the 50 was significant at the 5% level.

Focusing attention on the "explanatory" variables that proved significant on the first pass, a second model was constructed using only those 15 variables. The resulting model had an R^2 of 0.36 and the model coefficients of six of the "explanatory" (but completely unrelated) variables were significant at the 5% level. Given these findings, how can we be sure if the statistically significant variables we uncover in our own research via regression methods are truly explanatory or are merely the result of chance?

A partial answer may be found in an article by Gail Gong published in 1986 and reproduced in its entirety in Appendix 2.

Gail Gong was among the first, if not the first, student to have the bootstrap as the basis of her doctoral dissertation. Reading her article, reprinted here with the permission of the American Statistical Association, we learn the bootstrap can be an invaluable tool for model validation, a result we explore at greater length in the following chapter. We also learn not to take for granted the results of a stepwise regression.

Gong [1986] constructed a logistic regression model based on observations Peter Gregory made on 155 chronic hepatitis patients, 33 of whom died. The object of the model was to identify patients at high risk. In contrast to the computer simulations David Freedman performed, the 19 explanatory variables were real, not simulated, derived from medical histories, physical examinations, X-rays, liver function tests, and biopsies.

If one or more extreme values can influence the slope and intercept of a univariate regression line, think how much more impact, and how subtle the effect, these values might have on a curve drawn through 20-dimensional space.[1]

Gong's logistic regression models were constructed in two stages. At the first stage, each of the explanatory variables was evaluated on a univariate basis. Thirteen of these variables proved significant at the 5% level when applied to the original data. A forward multiple regression was applied to these thirteen variables and four were selected for use in the predictor equation.

When she took bootstrap samples from the 155 patients, the R^2 values of the final models associated with each individual bootstrap sample, varied widely. Not reported in this article, but far more important, is that while two of the original four predictor variables always appeared in the final model generated from a bootstrap sample of the patients, five other variables appeared in only *some* of the models.

We strongly urge you to adopt Dr. Gong's bootstrap approach to validating multi-variable models. Retain only those variables which appear consistently in the bootstrap regression models. Additional methods for model validation are described in Chapter 11.

Correcting for Confounding Variables

When your objective is to verify the association between predetermined explanatory variables and the response variable, multiple linear regression analysis permits you to provide for one or more confounding variables that could not be controlled otherwise.

GENERALIZED LINEAR MODELS

Today, most statistical software incorporates new advanced algorithms for the analysis of generalized linear models (GLMs)[2] and extensions to panel data settings including fixed-, random- and mixed-effects models, logistic-, Poisson, and negative-binomial regression, GEEs, and HLMs. These models take the form $Y = g^{-1}[\beta X] + \varepsilon$, where β is a vector of to-be-determined coefficients, X is a matrix of explanatory variables, and ε is a vector of identically distributed random variables. These variables may be normal, gamma, or Poisson depending on the specified variance of the GLM. The nature of the relationship between the outcome variable and the coefficients depend on the specified *link function g* of the GLM. Panel data models include the following:

Fixed Effects. An indicator variable for each subject is added and used to fit the model. Though often applied to the analysis of repeated measures,

---

[1] That's one dimension for risk of death, the dependent variable, and 19 for the explanatory variables.
[2] As first defined by Nelder and Wedderburn [1972].

this approach has bias that increases with the number of subjects. If data include a large number of subjects, the associated bias of the results makes this a very poor model choice.

Conditional Fixed Effects. These are applied in logistic regression, Poisson regression, and negative binomial regression. A sufficient statistic for the subject effect is used to derive a conditional likelihood such that the subject level effect is removed from the estimation.

While conditioning out the subject level effect in this manner is algebraically attractive, interpretation of model results must continue to be in terms of the conditional likelihood. This may be difficult and the analyst must be willing to alter the original scientific questions of interest to questions in terms of the conditional likelihood.

Questions always arise as to whether some function of the independent variable might be more appropriate to use than the independent variable itself. For example, suppose $X = Z^2$ where $E(Y|Z)$ satisfies the logistic equation; then $E(Y|X)$ does not.

Random Effects. The choice of a distribution for the random effect too often is driven by the need to find an analytic solution to the problem, rather than by any actual knowledge. If we assume a normally distributed random effect when the random effect is really Laplace, we will get the same point estimates (since both distributions have mean zero), but we will get different standard errors. We will not have any way of checking the approaches short of fitting both models.

If the true random effects distribution has a nonzero mean, then the misspecification is more troublesome as the point estimates of the fitted model are different from those that would be obtained from fitting the true model. Knowledge of the true random-effects distribution does not alter the interpretation of fitted model results. Instead, we are limited to discussing the relationship of the fitted parameters to those parameters we would obtain if we had access to the entire population of subjects, and we fit that population to the same fitted model. In other words, even given the knowledge of the true random effects distribution, we cannot easily compare fitted results to true parameters.

As discussed in Chapter 5 with respect to group-randomized trials, if the subjects are not independent (say, they all come from the same classroom), then the true random effect is actually larger. The attenuation of our fitted coefficient increases as a function of the number of supergroups containing our subjects as members; if classrooms are within schools and there is within school correlation, the attenuation is even greater.

GEE (Generalized Estimating Equation). Instead of trying to derive the estimating equation for GLM with correlated observations from a likelihood argument, the within subject correlation is introduced in the estimating equation itself. The correlation parameters are then nuisance parameters and can be estimated separately. (See also Hardin and Hilbe, 2003.)

Underlying the population-averaged GEE is the assumption that one is able to specify the correct correlation structure. If one hypothesizes an exchangeable correlation and the true correlation is time-dependent, the resulting regression coefficient estimator is inefficient. The naive variance estimates of the regression coefficients will then produce incorrect confidence intervals. Analysts specify a correlation structure to gain efficiency in the estimation of the regression coefficients, but typically calculate the sandwich estimate of variance to protect against misspecification of the correlation.[3] This variance estimator is more variable than the naive variance estimator, and many analysts do not pay adequate attention to the fact that the asymptotic properties depend on the number of subjects (not the total number of observations).

HLM. This includes hierarchical linear models, linear latent models, and others. While previous models are limited for the most part to a single effect, HLM allows more than one. Unfortunately, most commercially available software requires one to assume that each random effect is Gaussian with mean zero. The variance of each random effect must be estimated.

Mixed Models. These allow both linear and nonlinear mixed effects regression (with various links). They allow you to specify each level of repeated measures. Imagine: districts: schools: teachers: classes: students. In this description, each of the sublevels is within the previous level and we can hypothesize a fixed or random effect for each level. We also imagine that observations within same levels (any of these specific levels) are correlated.

The caveats revealed in this and the previous chapter apply to the GLMs. The most common sources of error are the use of an inappropriate or erroneous link function, the wrong choice of scale for an explanatory variable (for example, using x rather than $\log[x]$), neglecting important variables, and the use of an inappropriate error distribution when computing confidence intervals and p values. Firth [1991, pp. 74–77] should be consulted for a more detailed analysis of potential problems.

REPORTING YOUR RESULTS

In reporting the results of your modeling efforts you need to be explicit about the methods used, the assumptions made, the limitations on your

---

[3] See Hardin and Hilbe [2003, p. 28] for a more detailed explanation.

model's range of application, potential sources of bias, and the method of validation (see the following chapter). The section on "Limitations of the Logistic Regression" from Bent and Archfield [2002] is ideal in this regard:

"The logistic regression equation developed is applicable for stream sites with drainage areas between 0.02 and 7.00 mi^2 in the South Coastal Basin and between 0.14 and 8.94 mi^2 in the remainder of Massachusetts, because these were the smallest and largest drainage areas used in equation development for their respective areas." (The authors go on to subdivide the area.)

"The equation may not be reliable for losing reaches of streams, such as for streams that flow off area underlain by till or bedrock onto an area underlain by stratified-drift deposits (these areas are likely more prevalent where hillsides meet river valleys in central and western Massachusetts). At this juncture of the different underlying surficial deposit types, the stream can lose stream flow through its streambed. Generally, a losing stream reach occurs where the water table does not intersect the streambed in the channel (water table is below the streambed) during low-flow periods. In these reaches, the equation would tend to overestimate the probability of a stream flowing perennially at a site."

"The logistic regression equation may not be reliable in areas of Massachusetts where ground-water and surface-water drainage areas for a stream site differ." (The authors go on to provide examples of such areas.)

"In these areas, ground water can flow from one basin into another; therefore, in basins that have a larger ground-water contributing area than the surface-water drainage area the equation may underestimate the probability that stream is perennial. Conversely, in areas where the ground-water contributing area is less than the surface-water drainage area, the equation may overestimate the probability that a stream is perennial."

This report by Bent and Archfield also illustrates how data quality, selection, and measurement bias can restrict a model's applicability.

"The accuracy of the logistic regression equation is a function of the quality of the data used in its development. These data include the measured perennial or intermittent status of a stream site, the occurrence of unknown regulation above a site, and the measured basin characteristics.

"The measured perennial or intermittent status of stream sites in Massachusetts is based on information in the USGS NWIS database. Streamflow measured as less than 0.005 ft^3/s is rounded down to zero, so it is possible that several streamflow measurements reported as zero may have had flows less than 0.005 ft^3/s in the stream. This measurement would cause stream sites to be classified as intermittent when they actually are perennial."

"Additionally, of the stream sites selected from the NWIS database, 61 of 62 intermittent-stream sites and 89 of 89 perennial-stream sites were represented as perennial streams on USGS topographic maps; therefore, the Statewide database (sample) used in development of the equation may not be random, because stream sites often selected for streamflow measurements are represented as perennial streams on USGS topographic maps. Also, the drainage area of stream sites selected for streamflow measurements generally is greater than about $1.0\,mi^2$, which may result in the sample not being random."

"The observed perennial or intermittent status of stream sites in the South Coastal Basin database may also be biased, because the sites were measured during the summer of 1999. The summer of 1999 did not meet the definition of an extended drought; but monthly precipitation near the South Coastal Basin was less than 50 percent of average in April, less than 25 percent of average in June, about 75 percent of average in July (excluding one station), and about 50 percent of average in August (excluding one station). Additionally, Socolow and others (2000) reported streamflows and ground-water levels well below normal throughout most of Massachusetts during the summer of 1999. Consequently, stream sites classified as intermittent would have been omitted from the database had this period been classified as an extended drought. This climatic condition during the summer of 1999 could bias the logistic regression equation toward a lower probability of a stream site being considered perennial in the South Coastal Basin."

"Basin characteristics of the stream sites used in the logistic equation development are limited by the accuracy of the digital data layers used. In the future, digital data layers (such as hydrography, surficial geology, soils, DEMs, and land use) will be at lower scales, such as $1:5,000$ or $1:25,000$. This would improve the accuracy of the measured basin characteristics used as explanatory variables to predict the probability of a stream flowing perennially."

"For this study, the area of stratified-drift deposits and consequently the areal percentage of stratified-drift deposits included areas with sand and gravel, large sand, fine-grained, and floodplain alluvium deposits. Future studies would allow more specificity in testing the areal percentage of surficial deposits as explanatory variables. For example, the areal percentage of sand and gravel deposits may be an important explanatory variable for estimating the probability that a stream site is perennial. The accuracy of the logistic regression equation also may be improved with the testing of additional basin characteristics as explanatory variables. These explanatory variables could include areal percentage of wetlands (forested and non-forested), areal percentage of water bodies, areal percentage of forested land, areal percentage of urban land, or mean, minimum, and maximum basin elevation."

A CONJECTURE

A great deal of publicity has heralded the arrival of new and more powerful data mining methods—neural networks, CART, and dozens of unspecified proprietary algorithms. In our limited experience, none of these have lived up to expectations; see a report of our tribulations in Good [2001a, Section 7.6]. Most of the experts we've consulted have attributed this failure to the small size of our test data set, 400 observations each with 30 variables. In fact, many publishers of data mining software assert that their wares are designed solely for use with terra-bytes of information.

This observation has led to our putting our experience in the form of the following conjecture.

If m points are required to determine a univariate regression line with sufficient precision, then it will take at least m^n observations and perhaps $n!m^n$ observations to appropriately characterize and evaluate a model with n variables.

BUILDING A SUCCESSFUL MODEL

"Rome was not built in one day,"[4] nor was any reliable model. The only successful approach to modeling lies in a continuous cycle of hypothesis formulation–data gathering–hypothesis testing and estimation. How you go about it will depend on whether you are new to the field, have a small data set in hand, and are willing and prepared to gather more until the job is done, or you have access to databases containing hundreds of thousands of observations. The following prescription, while directly applicable to the latter case, can be readily modified to fit any situation.

1. A thorough literature search and an understanding of casual mechanisms is an essential prerequisite to any study. Don't let the software do your thinking for you.

2. Using a subset of the data selected at random, see which variables *appear* to be correlated with the dependent variable(s) of interest. (As noted in this and the preceding chapter, two unrelated variables may *appear* to be correlated by chance alone or as a result of confounding factors. For the same reasons, two closely related factors may fail to exhibit a statistically significant correlation.)

3. Using a second, distinct subset of the data selected at random, see which of the variables selected at the first stage still *appear* to be correlated with the dependent variable(s) of interest. Alternately, use the bootstrap method describe by Gong [1986] to see which variables are consistently selected for inclusion in the model.

4. Limit attention to one or two of the most significant predictor variables. Select a subset of the existing data which the remainder

---

[4] John Heywood, *Proverbes*, Part i, Chapter xi, 16th Century.

of the significant variables are (almost) constant. (Alternately, gather additional data for which the remainder of the significant variables are almost constant.) Decide on a generalized linear model form which best fits your knowledge of the causal relations among the few variables on which you are now focusing. (A standard multivariate linear regression may be viewed as just another form, albeit a particularly straightforward one, of generalized linear model.) Fit this model to the data.

5. Select a second subset of the existing data (or gather an additional data set) for which the remainder of the significant variables are (almost) equal to a second constant. For example, if only men were considered at stage four, then you should focus on women at this stage. Attempt to fit the model you derived at the preceding stage to these data.

6. By comparing the results obtained at stages four and five, you can determine whether to continue to ignore or to include variables previously excluded from the model. Only one or two additional variables should be added to the model at each iteration of steps 4 through 6.

7. Always validate your results as described in the next chapter.

If all this sounds like a lot of work, it is. It takes several years to develop sound models, even or despite the availability of lightning fast, multifunction statistical software. The most common error in statistics is to assume that statistical procedures can take the place of sustained effort.

TO LEARN MORE

Inflation of R^2 as a consequence of multiple tests also was considered by Rencher [1980].

Osborne and Waters [2002] review tests of the assumptions of multivariable regression. Harrell, Lee, and Mark [1996] review the effect of violation of assumptions on GLMs and suggest the use of the bootstrap for model validation. Hosmer and Lemeshow [2001] recommend the use of the bootstrap or some other validation procedure before accepting the results of a logistic regression.

Diagnostic procedures for use in determining an appropriate functional form are described by Mosteller and Tukey [1977], Therneau and Grambsch [2000], Hosmer and Lemeshow [2001], and Hardin and Hilbe [2003].

Chapter 11
Validation

". . . the simple idea of splitting a sample in two and then developing the hypothesis on the basis of one part and testing it on the remainder may perhaps be said to be one of the most seriously neglected ideas in statistics. If we measure the degree of neglect by the ratio of the number of cases where a method could help to the number of cases where it is actually used." G. A. Barnard in discussion following Stone [1974, p. 133].

Validate your models before drawing conclusions.

AS WE READ IN THE ARTICLES BY DAVID Freedman and Gail Gong reprinted in the Appendix absent a detailed knowledge of causal mechanisms, the results of a regression analysis are highly suspect. Freedman found highly significant correlations between totally independent variables. Gong resampled repeatedly from the data in hand and obtained a different set of significant variables each time.

A host of advertisements from new proprietary software claim an ability to uncover relationships previously hidden and to overcome the deficiencies of linear regression. But how can we determine whether or not such claims are true?

Good [2001a, Chapter 10] reports on one such claim from the maker of PolyAnalyst™. He took the 400 records, each of 31 variables, PolyAnalyst provided in an example dataset, split the data in half at random, and obtained completely discordant results with the two halves whether they were analyzed with PolyAnalyst, CART, or stepwise linear regression. This was yet another example of a spurious relationship that did not survive the validation process.

In this chapter we review the various methods of validation and provide guidelines for their application.

Common Errors in Statistics (and How to Avoid Them), by Phillip I. Good and James W. Hardin.
ISBN 0-471-46068-0   Copyright © 2003 John Wiley & Sons, Inc.

METHODS OF VALIDATION

Your choice of an appropriate methodology will depend upon your objectives and the stage of your investigation. Is the purpose of your model to predict—will there be an epidemic? to extrapolate—what might the climate have been like on the primitive Earth? or to elicit causal mechanisms—is development accelerating or decelerating? Which factors are responsible?

Are you still developing the model and selecting variables for inclusion, or are you in the process of estimating model coefficients?

There are three main approaches to validation:

1. **Independent verification (obtained by waiting until the future arrives or through the use of surrogate variables).**

2. **Splitting the sample (using one part for calibration, the other for verification).**

3. **Resampling (taking repeated samples from the original sample and refitting the model each time).**

Independent Verification

Independent verification is appropriate and preferable whatever the objectives of your model and whether selecting variables for inclusion or estimating model coefficients.

In soil, geologic, and economic studies, researchers often return to the original setting and take samples from points that have been bypassed on the original round. See, for example, Tsai et al. [2001].

In other studies, verification of the model's form and the choice of variables are obtained by attempting to fit the same model in a similar but distinct context.

For example, having successfully predicted an epidemic at one army base, one would then wish to see if a similar model might be applied at a second and third almost but not quite identical base.

Stockton and Meko [1983] reconstructed regional-average precipitation to A.D. 1700 in the Great Plains of the United States with multiple linear regression models calibrated on the period 1933–1977. They validated the reconstruction by comparing the reconstructed regional percentage-of-normal precipitation with single-station precipitation for stations with records extending back as far as the 1870s. Lack of appreciable drop in correlation between these single station records and the reconstruction from the calibration period to the earlier segment was taken as evidence for validation of the reconstructions.

Graumlich [1993] used a response-surface reconstruction method to reconstruct 1000 years of temperature and precipitation in the Sierra

Nevada. The calibration climatic data were 62 years of observed precipitation and temperature (1928–1989) at Giant Forest/Grant Grove. The model was validated by comparing the predictions with the 1873–1927 segments of three climate stations 90 km to the west in the San Joaquin Valley. The climatic records of these stations were highly correlated with those at Giant Forest/Grant Grove. Significant correlation of these long-term station records with the 1873–1927 part of the reconstruction was accepted as evidence of validation.

Independent verification can help discriminate among several models that appear to provide equally good fits to the data. Independent verification can be used in conjunction with either of the two other validation methods. For example, an automobile manufacturer was trying to forecast parts sales. After correcting for seasonal effects and long-term growth within each region, ARIMA techniques were used.[1] A series of best-fitting ARIMA models was derived, one model for each of the nine sales regions into which the sales territory had been divided. The nine models were quite different in nature. As the regional seasonal effects and long-term growth trends had been removed, a single ARIMA model applicable to all regions, albeit with differing coefficients, was more plausible. Accordingly, the ARIMA model that gave the best overall fit to all regions was utilized for prediction purposes.

Independent verification also can be obtained through the use of surrogate or proxy variables. For example, we may want to investigate past climates and test a model of the evolution of a regional or worldwide climate over time. We cannot go back directly to a period before direct measurements on temperature and rainfall were made, but we can observe the width of growth rings in long-lived trees or measure the amount of carbon dioxide in ice cores.

Sample Splitting

Splitting the sample into two parts—one for estimating the model parameters, the other for verification—is particularly appropriate for validating time series models where the emphasis is on prediction or reconstruction. If the observations form a time series, the more recent observations should be reserved for validation purposes. Otherwise, the data used for validation should be drawn at random from the entire sample.

Unfortunately, when we split the sample and use only a portion of it, the resulting estimates will be less precise.

Browne [1975] suggests we pool rather than split the sample if:

---

[1] For examples and discussion of AutoRegressive Integrated Moving Average processes, see Brockwell and Davis [1987].

(a) The predictor variables to be employed are specified beforehand (that is, we do not use the information in the sample to select them).

(b) The coefficient estimates obtained from a calibration sample drawn from a certain population are to be applied to other members of the same population.

The proportion to be set aside for validation purposes will depend upon the loss function. If both the goodness-of-fit error in the calibration sample and the prediction error in the validation sample are based on mean-squared error, Picard and Berk [1990] report that we can minimize their sum by using between one-fourth and one-third of the sample for validation purposes.

A compromise proposed by Moiser [1951] is worth revisiting: The original sample is split in half; regression variables and coefficients are selected independently for each of the subsamples; if they are more or less in agreement, then the two samples should be combined and the coefficients recalculated with greater precision.

A further proposal by Subrahmanyam [1972] to use weighted averages where there are differences strikes us as equivalent to painting over cracks left by the last earthquake. Such differences are a signal to probe deeper, to look into causal mechanisms, and to isolate influential observations that may, for reasons that need to be explored, be marching to a different drummer.

Resampling

We saw in the report of Gail Gong [1986], reproduced in Appendix B, that resampling methods such as the bootstrap may be used to validate our choice of variables to include in the model. As seen in last chapter, they may also be used to estimate the precision of our estimates.

But if we are to extrapolate successfully from our original sample to the population at large, then our original sample must bear a strong resemblance to that population. When only a single predictor variable is involved, a sample of 25 to 100 observations may suffice. But when we work with n variables simultaneously, sample sizes on the order of 25^n to 100^n may be required to adequately represent the full n-dimensional region.

Because of dependencies among the predictors, we can probably get by with several orders of magnitude fewer data points. But the fact remains that the sample size required for confidence in our validated predictions grows exponentially with the number of variables.

Five resampling techniques are in general use:

1. *K*-fold, in which we subdivide the data into *K* roughly equal-sized parts, then repeat the modeling process *K* times, leaving one section out each time for validation purposes.

2. Leave-one-out, an extreme example of K-fold, in which we subdivide into as many parts as there are observations. We leave one observation out of our classification procedure and use the remaining $n - 1$ observations as a training set. Repeating this procedure n times, omitting a different observation each time, we arrive at a figure for the number and percentage of observations classified correctly. A method that requires this much computation would have been unthinkable before the advent of inexpensive readily available high-speed computers. Today, at worst, we need step out for a cup of coffee while our desktop completes its efforts.

3. Jackknife, an obvious generalization of the leave-one-out approach, where the number left out can range from one observation to half the sample.

4. Delete-d, where we set aside a random percentage d of the observations for validation purposes, use the remaining $100 - d\%$ as a training set, and then average over 100 to 200 such independent random samples.

5. The bootstrap, which we have already considered at length in earlier chapters.

The correct choice among these methods in any given instance is still a matter of controversy (though any individual statistician will assure you the matter is quite settled). See, for example, Wu [1986] and the discussion following and Shao and Tu [1995].

Leave-one-out has the advantage of allowing us to study the influence of specific observations on the overall outcome.

Our own opinion is that if any of the above methods suggest that the model is unstable, the first step is to redefine the model over a more restricted range of the various variables. For example, with the data of Figure 9.3, we would advocate confining attention to observations for which the predictor (TNFAlpha) was less than 200.

If a more general model is desired, then many additional observations should be taken in underrepresented ranges. In the cited example, this would be values of TNFAlpha greater than 300.

MEASURES OF PREDICTIVE SUCCESS

Whatever method of validation is used, we need to have some measure of the success of the prediction procedure. One possibility is to use the sum of the losses in the calibration and the validation sample. Even this procedure contains an ambiguity that we need to resolve. Are we more concerned with minimizing the expected loss, the average loss, or the maximum loss?

One measure of goodness of fit of the model is $SSE = \Sigma(y_i - y_i^*)^2$, where y_i and y_i^* denote the ith observed value and the corresponding value obtained from the model. The smaller this sum of squares, the better the fit.

If the observations are independent, then

$$\sum (y_i - y_i^*)^2 = \sum (y_i - \bar{y})^2 - \sum (\bar{y} - y_i^*)^2.$$

The first sum on the right-hand side of the equation is the total sum of squares (SST). Most statistics software uses as a measure of fit $R^2 = 1 - SSE/SST$. The closer the value of R^2 is to 1, the better.

The automated entry of predictors into the regression equation using R^2 runs the risk of overfitting, because R^2 is guaranteed to increase with each predictor entering the model. To compensate, one may use the adjusted R^2

$$1 - [((n - i)(1 - R^2))/(n - p)]$$

where n is the number of observations used in fitting the model, p is the number of estimated regression coefficients, and i is an indicator variable that is 1 if the model includes an intercept and is 0 otherwise.

The adjusted R^2 has two major drawbacks according to Rencher and Pun [1980]:

1. **The adjustment algorithm assumes the predictors are independent; more often the predictors are correlated.**

2. **If the pool of potential predictors is large, multiple tests are performed, and R^2 is inflated in consequence; the standard algorithm for adjusted R^2 does not correct for this inflation.**

A preferable method of guarding against overfitting the regression model, proposed by Wilks [1995], is to use validation as a guide for stopping the entry of additional predictors. Overfitting is judged to begin when entry of an additional predictor fails to reduce the prediction error in the validation sample.

Mielke et al. [1997] propose the following measure of predictive accuracy for use with either a mean-square-deviation or a mean-absolute-deviation loss function:

$$M = 1 - \delta / \mu_\delta, \quad \text{where } \delta = \frac{1}{n} \sum_{i=1}^{n} |y_i - y_i^*| \quad \text{and} \quad \mu_\delta = \frac{1}{n^2} \sum_{i=1}^{n} \sum_{j=1}^{n} |y_i - y_j^*|.$$

Uncertainty in Predictions

Whatever measure is used, the degree of uncertainty in your predictions should be reported. Error bars are commonly used for this purpose.

The prediction error is larger when the predictor data are far from their calibration-period means, and vice versa. For simple linear regression, the standard error of the estimate s_e and standard error of prediction s_{y^*} are related as follows:

$$s_{y^*} = s_e \sqrt{\frac{(n+1)}{n} + (x_p - \bar{x})^2 \Big/ \sum_{i=1}^{n}(x_i - \bar{x})^2}$$

where n is the number of observations and x_i is the ith value of the predictor in the calibration sample, and x_p is the value of the predictor used for the prediction.

The relation between s_{y^*} and s_e is easily generalized to the multivariate case. In matrix terms, if $Y = AX + E$ and $y^* = AX_p$, then $s_{y^*}^2 = s_e^2\{1 + x_p^T(X^TX)^{-1}x_p\}$.

This equation is only applicable if the vector of predictors lies inside the multivariate cluster of observations on which the model was based. An important question is how "different" can the predictor data be from the values observed in the calibration period before the predictions are considered invalid.

LONG-TERM STABILITY

Time is a hidden dimension in most economic models. Many an airline has discovered to its detriment that what was an optimal price today leads to half-filled planes and markedly reduced profits tomorrow. A careful reading of the newspapers lets them know a competitor has slashed prices, but more advanced algorithms are needed to detect a slow shifting in tastes of prospective passengers. The public, tired of being treated no better than hogs,[2] turns to trains, personal automobiles, and teleconferencing.

An army base, used to a slow seasonal turnover in recruits, suddenly finds that all infirmary beds are occupied and the morning lineup for sick call stretches the length of a barracks.

To avoid a pound of cure:

- **Treat every model as tentative, best described, as any lawyer will advise you, as subject to change without notice.**
- **Monitor continuously.**

---

[2] Or somewhat worse, because hogs generally have a higher percentage of fresh air to breathe.

Most monitoring algorithms take the following form:

If the actual value exceeds some boundary value (the series mean, for example, or the series mean plus one standard deviation),

And if the actual value exceeds the predicted value for three observation periods in a row,

Sound the alarm (if the change like an epidemic is expected to be temporary in nature) or recalibrate the model.

TO LEARN MORE

Almost always, a model developed on one set of data will fail to fit a second independent sample nearly as well. Mielke et al. [1996] investigated the effects of sample size, type of regression model, and noise-to-signal ratio on the decrease or shrinkage in fit from the calibration to the validation data set.

For more on leave-one-out validation see Michaelsen [1987], Weisberg [1985], and Barnston and van den Dool [1993]. Camstra and Boomsma [1992] and Shao and Tu [1995] review the application of resampling in regression.

Watterson [1996] reviews the various measures of predictive accuracy.

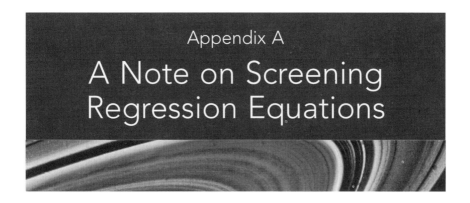

Appendix A

A Note on Screening Regression Equations

DAVID A. FREEDMAN*

Consider developing a regression model in a context where substantive theory is weak. To focus on an extreme case, suppose that in fact there is no relationship between the dependent variable and the explanatory variables. Even so, if there are many explanatory variables, the R^2 will be high. If explanatory variables with small t statistics are dropped and the equation refitted, the R^2 will stay high and the overall F will become highly significant. This is demonstrated by simulation and by asymptotic calculation.

KEY WORDS: Regression; Screening; R^2; F; Multiple testing.

1. INTRODUCTION

When regression equations are used in empirical work, the ratio of data points to parameters is often low; furthermore, variables with small coefficients are often dropped and the equations refitted without them. Some examples are discussed in Freedman (1981) and Freedman, Rothenberg, and Sutch (1982, 1983). Such practices can distort the significance levels of conventional statistical tests. The existence of this effect is well known, but its magnitude may come as a surprise, even to a hardened statistician. The object of the present note is to quantify this effect, both through

---

* David A. Freedman is Professor, Statistics Department, University of California, Berkeley, CA 94720. This research developed from a project supported by Dr. George Lady, of the former Office of Analysis Oversight and Access, Energy Information Administration, Department of Energy, Washington, D.C. I would like to thank David Brillinger, Peter Guttorp, George Lady, Thomas Permutt, and Thomas Rothenberg for their help. Reprinted with permission by *The American Statistician*.

Common Errors in Statistics (and How to Avoid Them), by Phillip I. Good and James W. Hardin.
ISBN 0-471-46068-0 Copyright © 2003 John Wiley & Sons, Inc.

simulation (Section 2) and through asymptotic calculation (Section 3). For another discussion, see Rencher and Pun (1980).

To help draw the conclusion explicitly, suppose an investigator seeks to predict a variable Y in terms of some large and indefinite list of explanatory variables $X_1, X_2, \ldots$. If the number of variables is comparable to the number of data points, and if the variables are only imperfectly correlated among themselves, then a very modest search procedure will produce an equation with a relatively small number of explanatory variables, most of which come in with significant coefficients, and a high significant R^2. This will be so even if Y is totally unrelated to the X's.

To sum up, in a world with a large number of unrelated variables and no clear a priori specifications, uncritical use of standard methods will lead to models that appear to have a lot of explanatory power. That is the main—and negative—message of the present note. Therefore, only the null hypothesis is considered here, and only the case where the number of variables is of the same order as the number of data points.

The present note is in the same spirit as the pretest literature. An early reference is Olshen (1973). However, there is a real difference in implementation: Olshen conditions on an F test being significant; the present note screens out the insignificant variables and refits the equation. Thus, Olshen has only one equation to deal with; the present note has two. The results of this note can also be differentiated from the theory of pretest estimators described in, for example, Judge and Bock (1978). To use the latter estimators, the investigator must decide a priori which coefficients may be set to zero; here, this decision is made on the basis of the data.

2. A SIMULATION

A matrix was created with 100 rows (data points) and 51 columns (variables). All the entries in this matrix were independent observations drawn from the standard normal distribution. In short, this matrix was pure noise. The 51st column was taken as the dependent variable Y in a regression equation; the first 50 columns were taken as the independent variables $X_1, \ldots, X_{50}$. By construction, then, Y was independent of the X's. Ideally, R^2 should have been insignificant, by the standard F test. Likewise, the regression coefficients should have been insignificant, by the standard t test.

These data were analyzed in two successive multiple regressions. In the first pass, Y was run on all 50 of the X's, with the following results:

- $R^2 = 0.50$, $P = 0.53$;
- 15 coefficients out of 50 were significant at the 25 percent level;
- 1 coefficient out of 50 was significant at the 5 percent level.

Only the 21 variables whose coefficients were significant at the 25 percent level were allowed to enter the equation on the second pass. The results were as follows:

- $R^2 = 0.36$, $P = 5 \times 10^{-4}$
- 14 coefficients out of 15 were significant at the 25 percent level;
- 6 coefficients out of 15 were significant at the 5 percent level.

The results from the second pass are misleading indeed, for they appear to demonstrate a definite relationship between Y and the X's, that is, between noise and noise. Graphical methods cannot help here; in effect, Y and the selected X's follow a jointly normal distribution conditioned on having significant t statistics. The simulation was done 10 times; the results are shown in Table 1. The 25 percent level was selected to represent an "exploratory" analysis; 5 percent for "confirmatory." The simulation was done in SAS on the UC Berkeley IBM 4341 by Mr. Thomas Permutt, on April 16, 1982.

3. SOME ASYMPTOTICS

An asymptotic calculation is helpful to explain the results of the simulation experiment. The Y and the X's are independent; condition X to be constant. There is no reason to treat the intercept separately since the Y's and X's all have expectation zero. Finally, suppose X has orthonormal columns. The resulting model is

$$Y = X\beta + \varepsilon \tag{1}$$

where Y is an $n \times 1$ random vector, X is a constant $n \times p$ matrix with orthonormal columns, where $p \geq p$, while β is a $p \times 1$ vector of parameters, and ε is an $n \times 1$ vector of independent normals, having mean 0 and common variance σ^2. In particular, the rank of X is p. All probabilities are computed assuming the null hypothesis that $\beta \equiv 0$. Suppose

$$n \to \infty \text{ and } p \to \infty \text{ so that } p/n \to \rho, \text{ where } 0 < \rho < 1. \tag{2}$$

Let R_n^2 be the square of the conventional multiple correlation coefficient, and F_n the conventional F statistic for testing the null hypothesis $\beta \equiv 0$. Under these conditions, the next proposition shows that R_n^2 will be essentially the ratio of the number p of variables to the number n of data points: the proof is deferred.

Proposition. Assume (1) and (2). Then

$$R_n^2 \to \rho \text{ and } F_n \to 1 \text{ in probability.} \tag{3}$$

TABLE 1. Simulation Results

Repetition	First Pass					Second Pass					
	R²	F	P[a]	#25%[b]	#5%[c]	R²	F	P × 10⁴	#p[b]	#25%	#5%
1	0.50	0.98	0.53	15	1	0.36	3.13	5	15	14	6
2	0.46	0.84	0.73	9	0	0.15	1.85	700	9	6	2
3	0.52	1.07	0.40	16	4	0.36	2.93	7	16	16	9
4	0.45	0.83	0.75	7	1	0.14	2.13	500	7	5	4
5	0.57	1.35	0.15	17	2	0.44	3.82	0.2	17	17	9
6	0.46	0.84	0.73	12	1	0.22	2.06	300	12	11	2
7	0.41	0.70	0.89	4	0	0.12	3.33	100	4	3	1
8	0.42	0.72	0.88	12	1	0.27	2.66	40	12	11	3
9	0.39	0.64	0.94	8	0	0.20	2.90	60	8	8	4
10	0.63	1.69	0.03	16	4	0.48	4.80	0.008	16	16	9

[a] P is the significance level of the F test, scaled up by 10^4 in the second pass.

[b] #25% is the number of variables whose coefficients are significant at the 25% level; only such variables are entered at the second pass; #p is the number of such variables, that is, the number of variables in the second pass regression, repeated for ease of reference.

[c] #5% is the number of variables whose coefficients are significant at the 5% level.

Note: The regressions are run without intercepts.

Now consider redoing the regression after dropping the columns of X that fail to achieve significance at level α. Here, $0 < \alpha < 1$ is fixed. Let $q_{n,\alpha}$ be the number of remaining columns. Let $R^2_{n,\alpha}$ be the square of the conventional multiple correlation in this second regression, and let $F_{n,\alpha}$ be the F statistic. These are to be computed by the standard formulas, that is, without any adjustment for the preliminary screening.

To estimate $R^2_{n,\alpha}$ and $F_{n,\alpha}$, the following will be helpful. Let Z be standard normal and $\Phi(z) = P\{|Z| > z\}$. Analytically,

$$\Phi(z) = \sqrt{\frac{2}{\pi}} \int_z^\infty \exp\left(-\frac{1}{2}u^2\right) du.$$

Choose λ so that $\Phi(\lambda) = \alpha$. Thus, λ is the cutoff for a two-tailed z test at level α. Let

$$g(z) = \int_{\{|z|>z\}} Z^2 < 1.$$

For $0 \leq z < \infty$, integration by parts shows

$$g(z) = \Phi(z) + \sqrt{\frac{2}{\pi}} z \exp\left(-\frac{1}{2}z^2\right). \tag{4}$$

Clearly,

$$E\{Z^2 | |Z| > z\} = g(z)/\Phi(z). \tag{5}$$

Then, as intuition demands,

$$E\{Z^2 | |Z| > z\} = 1 + \sqrt{\frac{2}{\pi}} \exp\left(-\frac{1}{2}z^2\right) \Big/ \Phi(z) > 1. \tag{6}$$

Let Z_λ be Z conditional on $|Z| > \lambda$. Put $z = \lambda$ in (5) and recall that $\Phi(\lambda) = \alpha$:

$$g(\lambda)/\alpha = E\{Z^2 | |Z| > \lambda\} = E\{Z_\lambda^2\} > 1 \tag{7}$$

Using (6) and further integration by parts.

$$\text{var}\{Z^2 | |Z| > z\} = 2 + v(z), \tag{8}$$

where

$$v(z)^2(z) = \sqrt{\frac{2}{\pi}} w(z) \exp\left(-\frac{1}{2}z^2\right) \Big/ \Phi(z)^2 \qquad (9)$$

and

$$w(z) = (z^3 + z)\Phi(z) - \sqrt{\frac{2}{\pi}}z^2 \exp\left(-\frac{1}{2}z^2\right).$$

In particular, v is continuous. Intuition suggests that v be positive. This fact will not be needed here, but it is true: see Diaconis and Freedman (1982, (3.15)–(3.16)).

Proposition. Assume (1) and (2). In probability: $q_{n,\alpha}/n \to \alpha\rho$ and $R^2_{n,\alpha} \to g(\lambda)$ and

$$F_{n,\alpha} \to \frac{g(\lambda)}{\alpha} \Big/ \frac{1 - g(\lambda)\rho}{1 - \alpha\rho}. \qquad (10)$$

In the second regression, the t statistic for testing whether a coefficient vanishes is asymptotically distributed as

$$Z_\lambda \sqrt{\frac{1 - \alpha\rho}{1 - g(\lambda)\rho}}.$$

These results may be interpreted as follows. The number of variables in the first-pass regression is $p = \rho n + o(n)$; the number in the second pass is $q_{n,\alpha} = \alpha\rho n + o(n)$. That is, as may be expected, α of the variables are significant at level α. Since $g(\lambda) < 1$, the R^2 in the second-pass regression is essentially the fraction $g(\lambda)$ of R^2 in the first pass. Likewise, $g(\lambda) > \alpha$, so the asymptotic value of the F statistic exceeds 1. Since the number of degrees of freedom is growing, off-scale P values will result. Finally, the real level of the t test may differ appreciably from the nominal level.

Example. Suppose $N = 100$ and $p = 50$, so $\rho = \frac{1}{2}$; and $\alpha = 0.25$ so $\lambda \doteq 1.15$. Then $g(\lambda) \doteq 0.72$, and $E\{Z^2| \, |Z| > \lambda\} \doteq 2.9$. In a regression with 50 explanatory variables and 100 data points, on the null hypothesis R^2 should be nearly $\frac{1}{2}$.

Next, run the regression again, keeping only the variables significant at the 25 percent level. The new R^2 should be around $g(\lambda) = 72$ percent of the original R^2. The new F statistic should be around

$$\frac{g(\lambda)}{\alpha} \Big/ \frac{1-g(\lambda)\rho}{1-\alpha\rho} \doteq 4.0.$$

The number of degrees of freedom should be around $\alpha\rho n \doteq 12$ in the numerator and $100 - 12 = 88$ in the denominator. (However, $q_{n,\alpha}$ is still quite variable, its standard deviation being about 3.) On this basis, a P value on the order of 10^{-4} may be anticipated.

What about the t tests? Take $\lambda' > \lambda$, corresponding to level $\alpha' < \alpha$. The nominal level for the test is α', but the real level is

$$\frac{1}{\alpha} P\left\{|Z| > \lambda' \sqrt{\frac{1-\alpha\rho}{1-g(\lambda)\rho}}\right\}.$$

Since $g(\lambda) > \alpha$, it follows that $1 - \alpha\rho > 1 - g(\lambda)\rho$. Keep $\alpha = 0.25$, so $\lambda \doteq 1.15$; take $\alpha' = 5$ percent, so $\lambda' = 1.96$; keep $\rho = \frac{1}{2}$. Now

$$\lambda' \sqrt{\frac{1-\alpha\rho}{1-g(\lambda)\rho}} = 2.3$$

and the real level is 9 percent. This concludes the example.

Turn now to the proofs. Without loss of generality, suppose the ith column of X has a 1 in the ith position and 0's everywhere else. Then

$$\hat{\beta}_i = Y_i \text{ for } i = 1, \dots, p,$$

and the sum of squares for error in the first-pass regression corresponding to the model (1) is

$$\sum_{i=p+1}^{n} Y_i^2.$$

Thus

$$R_n^2 = \sum_{i=1}^{p} Y_i^2 \Big/ \sum_{i=1}^{n} Y_i^2$$

and

$$F_n = \frac{1}{p} \sum_{i=1}^{p} Y_i^2 \Big/ \frac{1}{n-p} \sum_{i=p+1}^{n} Y_i^2.$$

Now (3) follows from the weak law of large numbers. Of course, $E(R_n^2)$ and var R_n are known: see Kendall and Stuart (1969).

To prove (10), the t statistic for testing $\beta_i = 0$ is Y_i/s_n, where

$$s_n^2 = \frac{1}{n-p} \sum_{j=p+1}^{n} Y_j^2.$$

Thus, column i of X enters the second regression iff $|Y_i/s_n| > t_{\alpha,n-p}$, the cutoff for a two-tailed t test at level α, with $n - p$ degrees of freedom.

In what follows, suppose without loss of generality that $\sigma^2 = 1$. Given s_n, the events

$$A_i = \{|Y_i| > t_{\alpha,n-p}s_n\}$$

are conditionally independent, with common conditional probability $\Phi(t_{\alpha,n-p}s_n)$. Of course, $t_{\alpha,n-p} \to \lambda$ and $s_n \to 1$; so this conditional probability converges to $\Phi(\lambda) = \alpha$. The number $q_{n,\alpha}$ of the events A_i that occur is therefore

$$\alpha p + o(p) = \alpha \rho n + o(n)$$

by (2). This can be verified in detail by computing the conditional expectation and variance.

Next, condition on s_n and $q_{n,\alpha} = q$ and on the identity of the q columns going into the second regression. By symmetry, suppose that it is columns 1 through q of X that enter the second regression. Then

$$R_{n,\alpha}^2 = \sum_{i=1}^{q} Y_i^2 \bigg/ \sum_{i=1}^{n} Y_i^2$$

and

$$F_{n,\alpha} = \frac{1}{q} \sum_{i=1}^{q} Y_i^2 \bigg/ \frac{1}{n-q} \sum_{i=q+1}^{n} Y_i^2.$$

Now $\sum_{i=1}^{n} Y_i^2 = n + o(n)$; and in the denominator of $F_{n,\alpha}$,

$$\sum_{i=q+1}^{n} Y_i^2 = \sum_{i=1}^{n} Y_i^2 - \sum_{i=1}^{q} Y_i^2.$$

It remains only to estimate $\sum_{i=1}^{q} Y_i^2$, to within $o(n)$. However, these Y_i's are conditionally independent, with common conditional distribution: they are distributed as Z given $|Z| > z_n$, where Z is $N(0, 1)$ and $z_n = t_{\alpha,n-p} \cdot s_n$. In view of (5), the conditional expectation of $\sum_{i=1}^{q} Y_i^2$ is

$$q_{n,\alpha} g(z_n)/\Phi(z_n).$$

But $q_{n,\alpha} = \alpha\rho n + o(n)$ and $z_n \to \lambda$. So the last display is, up to $o(n)$,

$$\alpha\rho n g(\lambda)/\alpha = g(\lambda)\rho n.$$

Likewise, the conditional variance of $\sum_{i=1}^q Y_i^2$ is $q_{n,\alpha}\{2 + v(z_n)\} = O(n)$; the conditional standard deviation is $O(\sqrt{n})$. Thus

$$\sum_{i=1}^q Y_i^2 = g(\lambda)\rho n + o(n),$$

$$\frac{1}{q}\sum_{i=1}^n Y_i^2 = g(\lambda)/\alpha + o(1),$$

$$\frac{1}{n-q}\sum_{i=q+1}^n Y_i^2 = \frac{1-g(\lambda)\rho}{1-\alpha\rho} + o(1).$$

This completes the argument for the convergence in probability. The assertion about the t statistic is easy to check, using the last display.

REFERENCES

Diaconis P; Freedman D. "On the Maximum Difference Between the Empirical and Expected Histograms for Sums," *Pacific Journal of Mathematics*, 1982; **100**:287–327.

Freedman D. "Some Pitfalls in Large-Scale Econometric Models: A Case Study," *University of Chicago Journal of Business*, 1981; **54**:479–500.

Freedman D; Rothenberg T; Sutch R. "A Review of a Residential Energy End Use Model," Technical Report No. 14, University of California, Berkeley, Dept. of Statistics, 1982.

—— "On Energy Policy Models," *Journal of Business and Economic Statistics*, 1983; **1**:24–32.

Judge G; Bock M. *The Statistical Implications of Pre-Test and Stein-Rule Estimators in Econometrics*, Amsterdam: North-Holland, 1978.

Kendall MG; Stuart A. *The Advanced Theory of Statistics*, London: Griffin, 1969.

Olshen RA. "The Conditional Level of the F-Test," *Journal of the American Statistical Association*, 1973; **68**, 692–698.

Rencher AC; Pun FC. "Inflation of R^2 in Best Subsets Regression," *Technometrics*, 1980; **22**:49–53.

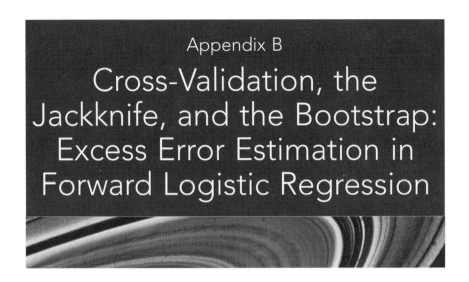

Appendix B

Cross-Validation, the Jackknife, and the Bootstrap: Excess Error Estimation in Forward Logistic Regression

GAIL GONG*

Given a prediction rule based on a set of patients, what is the probability of incorrectly predicting the outcome of a new patient? Call this probability the true error. An optimistic estimate is the apparent error, or the proportion of incorrect predictions on the original set of patients, and it is the goal of this article to study estimates of the excess error, or the difference between the true and apparent errors. I consider three estimates of the excess error: cross-validation, the jackknife, and the bootstrap. Using simulations and real data, the three estimates for a specific prediction rule are compared. When the prediction rule is allowed to be complicated, overfitting becomes a real danger, and excess error estimation becomes important. The prediction rule chosen here is moderately complicated, involving a variable-selection procedure based on forward logistic regression.

KEY WORDS: Prediction; Error rate estimation; Variables selection.

1. INTRODUCTION

A common goal in medical studies is prediction. Suppose we observe n patients, $x_1 = (t_1, y_1), \ldots, x_n = (t_n, y_n)$, where y_i is a binary variable indicating whether or not the ith patient dies of chronic hepatitis and t_i is a vector of explanatory variables describing various medical measurements

* Gail Gong is Assistant Professor, Department of Statistics, Carnegie-Mellon University, Pittsburgh, PA 15217.
Reprinted with permission by the American Statistical Association.

Common Errors in Statistics (and How to Avoid Them), by Phillip I. Good and James W. Hardin.
ISBN 0-471-46068-0   Copyright © 2003 John Wiley & Sons, Inc.

on the ith patient. These n patients are called the training sample. We apply a prediction rule η to the training sample x $= (x_1, \ldots, x_n)$ to form the realized prediction rule η_x. Given a new patient whose medical measurements are summarized by the vector t_0, we predict whether or not he will die of chronic hepatitis by $\eta_x(i_0)$, which takes on values "death" or "not death." Allowing the prediction rule to be complicated, perhaps including transforming and choosing from many variables and estimating parameters, we want to know: What is the error rate, or the probability of predicting a future observation incorrectly?

A possible estimate of the error rate is the proportion of errors that η_x makes when applied to the original observations $x_1, \ldots, x_n$. Because the same observations are used for both forming and assessing the prediction rule, this proportion, which I call the apparent error, underestimates the error rate.

To correct for this bias, we might use cross-validation, the jackknife, or the bootstrap for estimating excess errors (e.g., see Efron 1982). We study the performance of these three methods for a specific prediction rule. Excess error estimation is especially important when the training sample is small relative to the number of parameters requiring estimation, because the apparent error can be seriously biased. In the chronic hepatitis example, if the dimension of t_i is large relative to n, we might use a prediction rule that selects a subset of the variables that we hope are strong predictors. Specifically, I will consider a prediction rule based on forward logistic regression. I apply this prediction rule to some chronic hepatitis data collected at Stanford Hospital and to some simulated data. In the simulated data, I compare the performance of the three methods and find that cross-validation and the jackknife do not offer significant improvement over the apparent error, whereas the improvement given by the bootstrap is substantial.

A review of required definitions appears in Section 2. In Section 3, I discuss a prediction rule based on forward logistic regression and apply it to the chronic hepatitis data. In Sections 4 and 5, I apply the rule to simulated data. Section 6 concludes.

2. DEFINITIONS

I briefly review the definitions that will be used in later discussions. These definitions are essentially those given by Efron (1982). Let $x_1 = (t_1, y_1), \ldots, x_n = (t_n, y_n)$ be independent and identically distributed from an unknown distribution F, where t_i is a p-dimensional row vector of real-valued explanatory variables and y_i is a real-valued response. Let $\hat{F}$ be the empirical distribution function that puts mass $1/n$ at each point $x_1, \ldots, x_n$. We apply a prediction rule η to this training sample and form the realized prediction rule $\eta_{\hat{F}}(t_0)$. Let $Q(y_0, \eta_{\hat{F}}(t_0))$ be the criterion that scores the discrepancy between an observed value y_0 and its predicted value $\eta_{\hat{F}}(t_0)$. The form of both the prediction rule η and the criterion Q are given a priori. I define the *true error* of $\eta_{\hat{F}}$ to be the expected error that $\eta_{\hat{F}}$ makes on a new observation $x_0 = (t_0, y_0)$ from F,

$$q = q(\hat{F}, F) = E_{x_0 \sim F} Q(y_0, \eta_{\hat{F}}(t_0)).$$

In addition, I call the quantity

$$\hat{q}_{\text{app}} = q(\hat{F}, \hat{F}) = E_{x_0 \sim \hat{F}} Q(y_0, \eta_{\hat{F}}(t_0)) = \frac{1}{n} \sum_{i=1}^{n} Q(y_i, \eta_{\hat{F}}(t_i))$$

the *apparent error* of $\eta_{\hat{F}}$. The difference

$$R(\hat{F}, F) = q(\hat{F}, F) - q(\hat{F}, \hat{F})$$

is the *excess error* of $\eta_{\hat{F}}$. The *expected* excess error is

$$r = E_{\hat{F} \sim F} R(\hat{F}, F),$$

where the expectation is taken over $\hat{F}$, which is obtained from $x_1, \ldots, x_n$ generated by F. In Section 4, I will clarify the distinction between excess error and expected excess error. I will consider estimates of the expected excess error, although what we would rather have are estimates of the excess error.

I will consider three estimates (the bootstrap, the jackknife, and cross-validation) of the expected excess error. The bootstrap procedure for estimating $r = E_{\hat{F} \sim F} R(\hat{F}, F)$ replaces F with $\hat{F}$. Thus

$$\hat{r}_{\text{boot}} = E_{\hat{F}^* \sim \hat{F}} R(\hat{F}^*, \hat{F}),$$

where $\hat{F}^*$ is the empirical distribution function of a random sample $x_1^*, \ldots, x_n^*$ from $\hat{F}$. Since $\hat{F}$ is known, the expectation can in principle be calculated. The calculations are usually too complicated to perform analytically, however, so we resort to Monte Carlo methods.

1. Generate $x_1^*, \ldots, x_n^*$, a random sample from $\hat{F}$. Let $\hat{F}^*$ be the empirical distribution of $x_1^*, \ldots, x_n^*$.
2. Construct $\eta_{\hat{F}^*}$, the realized prediction rule based on $x_1^*, \ldots, x_n^*$.
3. Form

$$R^* = q(\hat{F}^*, \hat{F}) - q(\hat{F}^*, \hat{F}^*)$$

$$= \frac{1}{n}\sum_{i=1}^{n} Q(y_i, \eta_{\hat{F}^*}(t_i)) - \frac{1}{n}\sum_{i=1}^{n} Q(y_i^*, \eta_{\hat{F}^*}(t_i^*)) \qquad (2.1)$$

4. Repeat 1–3 a large number B times to get $R_1^*, \ldots, R_B^*$. The bootstrap estimate of expected excess error is

$$\hat{r}_{\text{boot}} = \frac{1}{B}\sum_{b=1}^{B} R_b^*.$$

See Efron (1982) for more details.

The jackknife estimate of expected excess error is

$$\hat{r}_{\text{jack}} = (n-1)(R_{(\cdot)} - \hat{R}),$$

where $\hat{F}^{(i)}$ is the empirical distribution function of $(x_1, \ldots, x_{i-1}, x_{i+1}, \ldots, x_n)$, and

$$R_{(i)} = R(\hat{F}^{(i)}, \hat{F}), \quad R_{(\cdot)} = \frac{1}{n}\sum_{i=1}^{n} R_{(i)}, \quad \hat{R} = R(\hat{F}, \hat{F}).$$

Efron (1982) showed that the jackknife estimate can be reexpressed as

$$\hat{r}_{\text{jack}} = \frac{1}{n}\sum_{i=1}^{n} Q(y_i, \eta_{\hat{F}^{(i)}}(t_i)) - \frac{1}{n}\sum_{i=1}^{n}\frac{1}{n}\sum_{j=1}^{n} Q(y_i, \eta_{\hat{F}^{(i)}}(t_i))$$

The cross-validation estimate of expected excess error is

$$\hat{r}_{\text{cross}} = \frac{1}{n}\sum_{i=1}^{n} Q(y_i, \eta_{\hat{F}^{(i)}}(t_i)) - \frac{1}{n}\sum_{i=1}^{n} Q(y_i, \eta_{\hat{F}}(t_i)).$$

Let the training sample omit patients one by one. For each omission, apply the prediction rule to the remaining sample and count the number (0 or 1) of errors that the realized prediction rule makes when it predicts the omitted patient. In total, we apply the prediction rule n times and predict the outcome of n patients. The proportion of errors made in these n predictions is the cross-validation estimate of the error rate and is the first term on the right-hand side. [Stone (1974) is a key reference on cross-validation and has a good historical account. Also see Geisser (1975).]

3. CHRONIC HEPATITIS: AN EXAMPLE

We now discuss a real prediction rule. From 1975 to 1980, Peter Gregory (personal communication, 1980) of Stanford Hospital observed $n = 155$ chronic hepatitis patients, of which 33 died from the disease. On each patient were recorded $p = 19$ covariates summarizing medical history, physical examinations, X rays, liver function tests, and biopsies. (Missing values were replaced by sample averages before further analysis of the data.) An effective prediction rule, based on these 19 covariates, was desired to identify future patients at high risk. Such patients require more aggressive treatment.

Gregory used a prediction rule based on forward logistic regression. We assume $x_1 = (t_1, y_1), \ldots, x_n = (t_n, y_n)$ are independent and identically distributed such that conditional on t_i, y_i is Bernoulli with probability of success $\theta(t_i)$, where logit $\theta(t_i) = \beta_0 + t_i\beta$, and where β is a column vector of p elements. If $(\hat{\beta}_0, \hat{\beta})$ is an estimate of (β_0, β), then $\hat{\theta}(t_0)$, such that logit $\hat{\theta}(t_0) = \hat{\beta}_0 + t_0\hat{\beta}$, is an estimate of $\theta(t_0)$. We predict death if the estimated probability $\hat{\theta}(t_0)$ of death were greater than $\frac{1}{2}$.:

$$\eta_{\hat{F}}(t_0) = 1 \quad \text{if } \hat{\theta}(t_0) \geq \frac{1}{2}, \text{ i.e., } \hat{\beta}_0 + t_0\hat{\beta} \geq 0$$

$$= 0 \quad \text{otherwise.} \tag{3.1}$$

Gregory's rule for estimating (β_0, β) consists of three steps.

1. Perform an initial screening of the variables by testing H_0: $\beta_j = 0$ in the simple logistic model, logit $\theta(t_0) = \beta + t_0\beta_j$, for $j = 1, \ldots, p$ *separately* at level $\alpha = 0.05$. Retain only those variables j for which the test is significant. Applied to Gregory's data, the initial screening retained 13 variables, 17, 12, 14, 11, 13, 19, 6, 5, 18, 10, 1, 4, 2, in increasing order of p-values.

2. To the variables that were retained in the initial screening, apply forward logistic regression that adds variables one at a time in the following way. Assume variables $j_1, j_2, \ldots, j_{P_1}$ are already added to the model. For each remaining j, test H_0: $\beta_j = 0$ in the linear logistic model that contains variables $j_1, j_2, \ldots, j_{P_1}, j$ together with the intercept. Rao's (1973, pp. 417–420) efficient score test requires calculating the maximum likelihood estimate only under H_0. If the most significant variable is significant at $\alpha = 0.05$, we add that variable to the model as variable j_{P_1+1} and start again. If none of the remaining variables is significant at $\alpha = 0.05$, we stop. From the aforementioned 13 variables, forward logistic regression applied to Gregory's data chose four variables (17, 11, 14, 2) that are, respectively, albumin, spiders, bilirubin, and sex.

3. Let $(\hat{\beta}_0, \hat{\beta})$ be the maximum likelihood estimate based on the linear logistic model consisting of the variables chosen by forward logistic regression together with the intercept. On Gregory's data, it turned out that

$$\left(\hat{\beta}_0, \hat{\beta}_{17}, \hat{\beta}_{11}, \hat{\beta}_{14}, \hat{\beta}_2 \right) = \left(12.17, -1.83, -1.58, 0.56, -5.17 \right)$$

The realization $\eta_{\hat{F}}$ of Gregory's rule on his 155 chronic hepatitis patients predicts that a new patient with covariate vector t_0 will die if his predicted probability of death $\hat{\theta}(t_0)$ is greater than $\frac{1}{2}$; that is,

$$\text{logit } \hat{\theta}(t_0) = 12.17 - 1.83t_{0.17} - 1.58t_{0.11} + 0.56t_{0,14} - 5.17t_{0,3} \geq \quad (3.2)$$

For the dichotomous problem, we use the criterion

$$Q(y, \eta) = 1 \quad \text{if } y \neq \eta,$$
$$= 0 \quad \text{otherwise}$$

The apparent error is $\hat{q}_{\text{app}} = 0.136$. Figure 1 shows a histogram of $B = 400$ bootstrap replications of $R^* = R(\hat{F}^*, \hat{F})$. Recall that each R^* was calculated using (2.1), where $\eta_{\hat{F}^*}$ is the realization of Gregory's rule on the bootstrap sample $x_1^*, \ldots, x_n^*$. The bootstrap estimate of expected excess error was

$$\hat{r}_{\text{boot}} \sim \frac{1}{B} \sum_{b=1}^{B} R_b^* = 0.039.$$

The jackknife and cross-validation estimates were calculated to be

$$\hat{r}_{\text{jack}} = 0.023, \quad \hat{r}_{\text{cross}} = 0.019.$$

Adding expected excess error estimates to the apparent error gives bias-corrected estimates of the error:

$$\hat{q}_{\text{boot}} = 0.175, \quad \hat{q}_{\text{jack}} = 0.159, \quad \hat{q}_{\text{cross}} = 0.145.$$

All three estimates require substantial computing time. FORTRAN programs for performing the preceding calculations and the ones in the following section were developed on a PDP-11/34 minicomputer. The cross-validation and jackknife estimates were computed in $1\frac{1}{2}$ hours, whereas the 400 bootstrap replications required just under 6 hours. Computers are becoming faster and cheaper, however, and even now it is possible to compute these estimates on very complicated prediction rules, such as Gregory's rule.

Are $B = 400$ bootstrap replications enough? Notice that $R_1^*, \ldots, R_B^*$ is a random sample from a population with mean

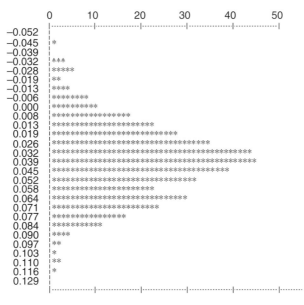

```
     0      10      20      30      40      50
     |········|········|········|········|········|
-0.052  |
-0.045  |*
-0.039  |
-0.032  |***
-0.028  |*****
-0.019  |**
-0.013  |****
-0.006  |********
 0.000  |**********
 0.008  |*****************
 0.013  |**********************
 0.019  |*************************
 0.026  |********************************
 0.032  |*******************************************
 0.039  |***********************************************
 0.045  |******************************************
 0.052  |*********************************
 0.058  |**********************
 0.064  |******************************
 0.071  |***********************
 0.077  |*****************
 0.084  |*************
 0.090  |****
 0.097  |**
 0.103  |*
 0.110  |**
 0.116  |*
 0.129  |
        |········|········|········|········|········|········|
```

FIGURE 1 Histogram of Bootstrap Replications for Gregory's Rule. The histogram summarizes the 400 bootstrap replications of R* that are used in estimating the expected excess error of Gregory's rule for predicting death in chronic hepatitis. Values of R* range from –0.045 to 0.116, with mean 0.039, standard deviation 0.027, and quantiles $R^*_{(.05)} = -0.006$ and $R^*_{(.05)} = 0.084$.

$$E_{\hat{F}^* \sim \hat{F}} R(\hat{F}^*, \hat{F}) = \hat{r}_{\text{boot}} = \hat{r}_\infty,$$

and variance, say, σ^2. Figure 1 shows that this population is close to normal, so

$$|\hat{r}_{400} - \hat{r}_\infty| \le 2\sigma / 400^{1/2},$$

with high probability. Approximating σ^2 with

$$\hat{\sigma}^2_{400} = \frac{1}{400-1} \sum_{b-1}^{400} [R^*_b - \hat{r}_{400}]^2 = (0.027)^2$$

gives

$$|\hat{r}_{400} - \hat{r}_\infty| \le 2(0.027) / 400^{1/2} = 0.0027;$$

so with high probability, $\hat{r}_{400}$ is within 0.0027 of $\hat{r}_\infty = \hat{r}_{\text{boot}}$.

Before leaving the chronic hepatitis data, I mention that other prediction rules might be used. Examples include more complicated forms of variable selection such as best subset regression and alternative models such as discriminant analysis. Friedman (1977) applied recursive partitioning to these data to obtain a binary-decision tree. I chose to focus attention on the rule based on forward logistic regression because it is the rule actually proposed and used by Gregory, the experimenter. The question of choosing an optimal prediction rule was not my goal.

4. THE PERFORMANCE OF CROSS-VALIDATION, THE JACKKNIFE, AND THE BOOTSTRAP IN SIMULATIONS

In the previous section we saw the cross-validation, jackknife, and bootstrap estimates of expected excess error for Gregory's rule. These estimates give bias corrections to the apparent error. Do these corrections offer real improvements? Introduce the "estimators" $\hat{r}_{app} \equiv 0$, the zero-correction estimate corresponding to the apparent error, and $\hat{r}_{ideal} \equiv E(R)$, the best constant estimate if we knew the expected excess error $E(R)$. To compare $\hat{r}_{cross}$, $\hat{r}_{jack}$, $\hat{r}_{boot}$ against these worst and best cases $\hat{r}_{app}$ and $\hat{r}_{ideal}$, we perform some simulations.

To judge the performance of estimators in the simulations, we use two criteria:

$$\mathrm{RMSE}_1\left(\hat{R}\right) = \left(E\left[\hat{R} - R\right]^2\right)^{1/2},$$

the root mean squared error (RMSE) about the excess error, and

$$\mathrm{RMSE}_2\left(\hat{R}\right) = \left(E\left[\hat{R} - E(R)\right]^2\right)^{1/2},$$

the root mean squared error about the expected excess error. Notice that since

$$E\left[\hat{R} - R\right]^2 = E\left[\left(\hat{q}_{app} + \hat{R}\right) - \left(\hat{q}_{app} + \hat{R}\right)\right]^2,$$

$\mathrm{RMSE}_1(\hat{R})$ also measures the performance of the bias-corrected estimate $\hat{q}_{app} + \hat{R}$ as an estimator of the true error $\hat{q}_{app} + R$.

I pause to clarify the distinction between excess error and expected excess error. In the chronic hepatitis problem, the training sample that Gregory observed led to a particular realization (3.2). The excess error is the difference between the true and apparent error of the realized rule $\eta_{\hat{F}}$ based on this training sample. The expected excess error averages the excess error over the many training samples that Gregory might have observed and, therefore, over many realizations of this prediction rule. Because $\hat{r}_{\text{cross}}$, $\hat{r}_{\text{jack}}$, $\hat{r}_{\text{boot}}$ average over many realizations, they are, strictly speaking, estimates of the expected excess error. Gregory, however, would much rather know the excess error of his particular realization.

It is perhaps unfair to think of $\hat{r}_{\text{cross}}$, $\hat{r}_{\text{jack}}$, $\hat{r}_{\text{boot}}$ as estimators of the excess error. A simple analogy may be helpful. Suppose X is an observation from the distribution F_ζ, and $T(X)$ estimates ζ. The bias is the expected difference $E[T(X) - \zeta]$ and is analogous to the expected excess error. The difference $T(X) - \zeta$ is analogous to the excess error. Getting a good estimate of the bias is sometimes possible, but getting a good estimate of the difference $T(X) - \zeta$ would be equivalent to knowing ζ.

In the simulations, the underlying model was the logistic model that assumes $x_1 = (t_1, y_1), \ldots, x_n = (t_n, y_n)$ are independent and identically distributed such that y_i conditional on t_i is Bernoulli with probability of success $\theta(t_i)$, where

$$\text{logit } \theta(t_i) = \beta_0 + t_i\beta, \tag{4.1}$$

where $t_i = (t_{i1}, \ldots, t_{ip})$ is p-variate normal with zero mean and a specified covariance structure Σ.

I performed two sets of simulations. In the first set (simulations 1.1, 1.2, 1.3) I let the sample sizes be, respectively, $n = 20, 40, 60$; the dimension of t_i be $p = 4$; and

$$\Sigma = \begin{pmatrix} 1 & 0 & 0 & 0 \\ 0 & 0 & \tau & 0 \\ 0 & \tau & 1 & 0 \\ 0 & 0 & 0 & 1 \end{pmatrix}, \quad \beta_0 = 0, \quad \beta = \begin{pmatrix} 1 \\ 2 \\ 0 \\ 0 \end{pmatrix}, \tag{4.2}$$

where $\tau = 0.80$. We would expect a good prediction rule to choose variables t_1 and t_2, and due to the correlation between variables t_2 and t_3, a prediction rule choosing t_1 and t_3 would probably not be too bad. In the second set of simulations (simulations 2.1, 2.2, 2.3, the sample sizes were again $n = 20, 40, 60$; the dimension of t_i was increased to $p = 6$; and

$$\Sigma = \begin{pmatrix} 1 & 0 & 0 & 0 & 0 & 0 \\ 0 & 1 & 0 & 0 & 0 & 0 \\ 0 & 0 & 1 & 0 & 0 & 0 \\ 0 & 0 & 0 & 1 & \tau & 0 \\ 0 & 0 & 0 & \tau & 1 & 0 \\ 0 & 0 & 0 & 0 & 0 & 1 \end{pmatrix}, \quad \beta_0 = 0, \quad \beta = \begin{pmatrix} 1 \\ 1 \\ 1 \\ 2 \\ 0 \\ 0 \end{pmatrix}. \tag{4.3}$$

TABLE 1 The Results of 400 Experiments of Simulation 1.1

$\hat{R}$	$E(\hat{R})$	$SD(\hat{R})$	$RMSE_1(\hat{R})$	$RMSE_2(\hat{R})$
apparent	0.0000	0.0000	0.1354	0.1006
cross	0.1039	0.1060	0.1381	0.1060
jack	0.0951	0.0864	0.1274	0.0865
boot	0.0786	0.0252	0.1078	0.0334
ideal	0.1006	0.0000	0.0906	0.0000

Note: $RMSE_1$ is the root mean squared error about the true excess, and $RMSE_2$ is that about the expected excess error. The expected excess error is $E(\hat{R})$ for ideal.

Each of the six simulations consisted of 400 experiments. The results of all 400 experiments of simulation 1.1 are summarized in Table 1. In each experiment, we estimate the excess error R by evaluating the realized prediction rule on a large number (5,000) of new observations. We estimate the expected excess error by the sample average of the excess errors in the 400 experiments. To compare the three estimators, I first remark that in the 400 experiments, the bootstrap estimate was closest to the true excess error 210 times. From Table 1 we see that since

$$E(\hat{r}_{\mathrm{cross}}) = 0.1039, \quad E(\hat{r}_{\mathrm{jack}}) = 0.0951, \quad E(R) = 0.1006$$

are all close, $\hat{r}_{\mathrm{cross}}$ and $\hat{r}_{\mathrm{jack}}$ are nearly unbiased estimates of the expected excess error $E(R)$, whereas $\hat{r}_{\mathrm{boot}}$ with expectation $E(\hat{r}_{\mathrm{boot}}) = 0.0786$ is biased downwards. [Actually, since we are using the sample averages of the excess errors in 400 experiments as estimates of the expected excess errors, we are more correct in saying that a 95% confidence interval for $E(\hat{r}_{\mathrm{cross}})$ is (0.0935), 0.1143), which contains $E(R)$, and a 95% confidence interval for $E(\hat{r}_{\mathrm{jack}})$ is (0.0866, 1036), which also contains $E(R)$. On the other hand, a 95% confidence interval for $E(\hat{r}_{\mathrm{boot}})$ is (0.0761, 0.0811), which does not contain $E(R)$.] However, $\hat{r}_{\mathrm{corss}}$ and $\hat{r}_{\mathrm{jack}}$ have enormous standard deviations, 0.1060 and 0.0864, respectively, compared to 0.0252, the standard deviation of $\hat{r}_{\mathrm{boot}}$. From the column for $RMSE_1$,

$$RMSE_1(\hat{r}_{\mathrm{ideal}}) < RMSE_1(\hat{r}_{\mathrm{boot}}) < RMSE_1(\hat{r}_{\mathrm{app}}) \sim RMSE_1(\hat{r}_{\mathrm{cross}}) \sim RMSE_1(\hat{r}_{\mathrm{jack}}),$$

with $RMSE_1(\hat{r}_{\mathrm{boot}})$ being about one-third of the distance between $RMSE_1(\hat{r}_{\mathrm{ideal}})$ and $RMSE_1(\hat{r}_{\mathrm{app}})$. The same ordering holds for $RMSE_2$.

Recall that simulations 1.1, 1.2, and 1.3 had the same underlying distribution but differing sample sizes, $n = 20$, 40, and 60. As sample size increased, the expected excess error decreased, as did the mean squared error of the apparent error. We observed a similar pattern in simulations 2.1, 2.2, and 2.3, where the sample sizes were again $n = 20$, 40, and 60,

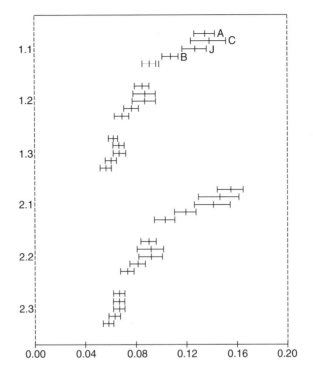

FIGURE 2 95% (nonsimultaneous) Confidence Intervals for $RMSE_1$. In each set of simulations, there are five confidence intervals for, respectively, apparent (A), cross-validation (C), jackknife (J), bootstrap (B), and ideal (1) estimates of the excess error. Each confidence interval is indicated by — —. The middle vertical bar in each confidence interval represents the value of the estimate.

and the dimension of t_i was increased to $p = 6$ and Σ, β_0, and β given in (4.3). For larger sample sizes, bias corrections to the apparent error became less important. It is still interesting, however, to compare mean squared errors. For all six simulations, I plot $RMSE_1$'s in Figure 2 and $RMSE_2$'s in Figure 3. It is interersting to note that the ordering noticed in simulation 1.1 of the root mean squared error of the five estimates also held in the other five simulations. That is,

$$RMSE_1(\hat{r}_{app}) \sim RMSE_1(\hat{r}_{cross}) \sim RMSE_1(\hat{r}_{jack}),$$

and $RMSE_1(\hat{r}_{boot})$ is about one-third of the distance between $RMSE_1(\hat{r}_{ideal})$ and $RMSE_1(\hat{r}_{app})$. Similar remarks hold for $RMSE_2$. Cross-validation and the jackknife offer no improvement over the apparent error, whereas the improvement given by the bootstrap is substantial.

The superiority of the bootstrap over cross-validation has been observed in other problems. Efron (1983) discussed estimates of excess error and

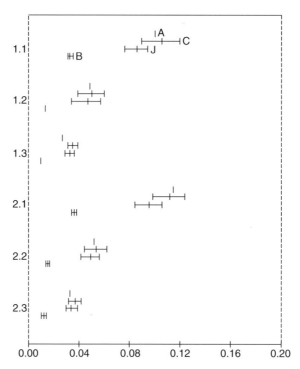

FIGURE 3 95% (nonsimultaneous) Confidence Intervals for $RMSE_2$. In each set of simulations, there are four confidence intervals for, respectively, apparent (A), cross-validation (C), jackknife (J), and bootstrap (B) estimates of the expected excess error. Notice that $\hat{r}_{app} \equiv 0$, so $RMSE_2(\hat{r}_{app})$ is the expected excess error, a constant; the "confidence interval" for $RMSE_2(\hat{r}_{app})$ is a single value, indicated by a single bar. In addition, $RMSE_2(\hat{r}_{ideal}) = 0$ and its confidence intervals are not shown. Some of the bootstrap confidence intervals are so small that they are indistinguishable from single bars.

performed several simulations with a flavor similar to mine. I report on only one of his simulations here. When the prediction rule is the usual Fisher discriminant and the training sample consists of 14 observations that are equally likely from $\mathfrak{N}((-\frac{1}{2}, 0), I)$ or $\mathfrak{N}((+\frac{1}{2}, 0), I)$, then the $RMSE_1$ of apparent, cross-validation, bootstrap, and ideal estimates are, respectively, 0.149, 0.144, 0.134, and 0.114. Notice that the $RMSE_1$'s of cross-validation and apparent estimates are close, whereas the $RMSE_1$ of the bootstrap estimate is about halfway between that of the ideal and apparent estimates.

In the remainder of this section, I discuss the sufficiency of the number of bootstrap replications and the number of experiments.

Throughout the simulations, I used $B = 100$ bootstrap replications for each experiment. Denote

$$\hat{r}_B = \frac{1}{B} \sum_{b=1}^{B} R_b^*, \quad M(B) = \text{MSE}_1(\hat{r}_B).$$

Using a component-of-variance calculation (Gong 1982), for Simulation 1.1

$$M^{1/2}(\infty) = 0.1070 \sim 0.1078 = M^{1/2}(100);$$

so if we are interested in comparing root mean squared errors about the excess error, we need not perform more than $B = 100$ bootstrap replications.

In each simulation, I included 400 experiments and therefore used the approximation

$$\text{MSE}_1(\hat{r}) \equiv E[\hat{r} - R]^2 \sim \frac{1}{400} \sum_{e=11}^{400} [\hat{r}_e - R_e]^2,$$

where $\hat{r}_e$ and R_e are the estimate and true excess of the eth experiment. Figure 2 and 3 show 95% nonsimultaneous confidence intervals for RMSE_1's and RMSE_2's. Shorter intervals for RMSE_1's would be preferable, but obtaining them would be time-consuming. Four hundred experiments of simulation 1.1 with $p = 4$, $n = 20$, and $B = 100$ took 16 computer hours on the PDP-11/34 minicomputer, whereas 400 experiments of simulation 2.3 with $p = 6$, $n = 60$, and $B = 100$ took 72 hours. Halving the length of the confidence intervals in Figures 2 and 3 would require four times the number of experiments and four times the computer time. On the other hand, for each simulation in Figure 3, the confidence interval for $\text{RMSE}_2(\hat{r}_{\text{ideal}})$ is disjoint from that of $\text{RMSE}_2(\hat{r}_{\text{boot}})$, and both and disjoint from the confidence intervals for $\text{RMSE}_2(\hat{r}_{\text{jack}})$, $\text{RMSE}_2(\hat{r}_{\text{cross}})$, and $\text{RMSE}_2(\hat{r}_{\text{app}})$. Thus, for RMSE_2, we can convincingly argue that the number of experiments is sufficient.

5. THE RELATIONSHIP BETWEEN CROSS-VALIDATION AND THE JACKKNIFE

Efron (1982) conjectured that the cross-validation and jackknife estimates of excess error are asymptotically close. Gong (1982) proved Efron's conjecture. Unfortunately, the regularity conditions stated there do not hold for Gregory's rule. The conjecture seems to hold for Gregory's rule, however, as evidenced in Figure 4, a scatterplot of the jackknife and cross-validation estimates of the first 100 experiments of simulation 1.1. The plot shows points hugging the 45° line, whereas a scatterplot of the bootstrap and cross-validation exhibits no such behavior.

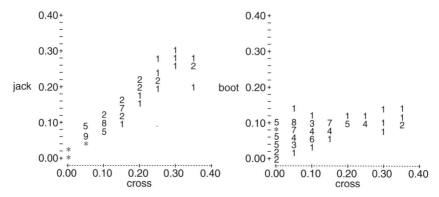

FIGURE 4 Scatterplots to Compare $\hat{r}_{corss}$, $\hat{r}_{jack}$ and $\hat{r}_{boot}$. The scatterplots summarize the relationships among the three estimates for the first 100 experiments of simulation 1.1. The numerals indicate the number of observations; * indicates greater than 9.

6. CONCLUSIONS

Because complicated prediction rules depend intricately on the data and thus have grossly optimistic apparent errors, error rate estimation for complicated prediction rules is an important problem. Cross-validation is a time-honored tool for improving the apparent error. This article compares cross-validation with two other methods, the jackknife and the bootstrap. With the help of increasingly available computer power, all three methods are easily applied to Gregory's complicated rule for predicting the outcome of chronic hepatitis. Simulations suggest that whereas the jackknife and cross-validation do not offer significant improvement over the apparent error, the bootstrap shows substantial gain.

REFERENCES

Efron B. *The Jackknife, the Bootstrap, and Other Resampling Plans*, Philadelphia: Society for Industrial and Applied Mathematics 1982.

—— "Estimating the Error Rate of a Prediction Rule: Improvements on Cross-Validation," *Journal of the American Statistical Association*, 1983; **78**:316–331.

Friedman JR. "A Recursive Partitioning Decision Rule for Nonparametric Classification," *IEEE Transactions on Computers*, C-26, 1977; 404–408.

Geisser S. "The Predictive Sample Reuse Method With Applications," *Journal of the American Statistical Association*, 1975; **70**:320–328.

Gong G. "Cross-Validation, the Jackknife, and the Bootstrap: Excess Error Estimation in Forward Logistic Regression," unpublished Ph.D. thesis, Stanford University 1982.

Rao CR. *Linear Statistical Inference and Its Applications*, New York: John Wiley 1973.

Stone M. "Cross-Validatory Choice and Assessment of Statistical Predictions," *Journal of the Royal Statistical Society*, 1974; **36**:111–147.

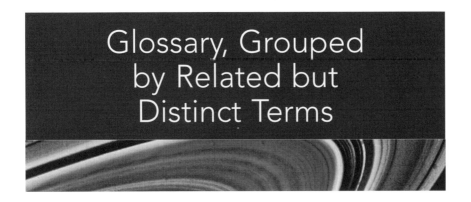

Glossary, Grouped by Related but Distinct Terms

ACCURACY AND PRECISION

An *accurate* estimate is close to the estimated quantity. A *precise* interval estimate is a narrow one. Precise measurements made with a dozen or more decimal places may still not be accurate.

DETERMINISTIC AND STOCHASTIC

A phenomenon is *deterministic* when its outcome is inevitable and all observations will take specific value.[1] A phenomenon is *stochastic* when its outcome may take different values in accordance with some probability distribution.

DICHOTOMOUS, CATEGORICAL, ORDINAL, METRIC DATA

Dichotomous data have two values and take the form "yes or no," "got better or got worse."

Categorical data have two or more categories such as yes, no, and undecided. Categorical data may be ordered (opposed, indifferent, in favor) or unordered (dichotomous, categorical, ordinal, metric).

Preferences can be placed on an ordered or *ordinal* scale such as strongly opposed, opposed, indifferent, in favor, strongly in favor.

Metric data can be placed on a scale that permits meaningful subtraction; for example, while "in favor" minus "indifferent" may not be meaningful, 35.6 pounds minus 30.2 pounds is.

Metric data can be grouped so as to evaluate them by statistical methods applicable to categorical or ordinal data. But to do so would be

---

[1] These observations may be subject to measurement error.

Common Errors in Statistics (and How to Avoid Them), by Phillip I. Good and James W. Hardin. ISBN 0-471-46068-0   Copyright © 2003 John Wiley & Sons, Inc.

to throw away information and also reduce the power of any tests and the precision of any estimates.

DISTRIBUTION, CUMULATIVE DISTRIBUTION, EMPIRICAL DISTRIBUTION, LIMITING DISTRIBUTION

Suppose we were able to examine all the items in a population and record a value for each one to obtain a *distribution* of values. The *cumulative distribution function* of the population $F[x]$ denotes the probability that an item selected at random from this population will have a value less than or equal to x. $0 \leq F[x] \leq 1$. Also, if $x < y$, then $F[x] \leq F[y]$.

The *empirical distribution,* usually represented in the form of a cumulative frequency polygon or a bar plot, is the distribution of values observed in a sample taken from a population. If $F_n[x]$ denotes the cumulative distribution of observations in a sample of size n, then as the size of the sample increases we have $F_n[x] \rightarrow F[x]$.

The *limiting distribution* for very large samples of a sample statistic such as the mean or the number of events in a large number of very small intervals often tends to a distribution of known form such as the Gaussian for the mean or the Poisson for the number of events.

Be wary of choosing a statistical procedures which is optimal only for a limiting distribution and not when applied to a small sample. For a small sample, the empirical distribution may be a better guide.

HYPOTHESIS, NULL HYPOTHESIS, ALTERNATIVE

The dictionary definition of a *hypothesis* is a proposition, or set of propositions, put forth as an explanation for certain phenomena.

For statisticians, a *simple hypothesis* would be that the distribution from which an observation is drawn takes a specific form. For example, $F[x]$ is $N(0,1)$. In the majority of cases, a statistical hypothesis will be *compound* rather than simple—for example, that the distribution from which an observation is drawn has a mean of zero.

Often, it is more convenient to test a *null hypothesis*—for example, that there is no or null difference between the parameters of two populations.

There is no point in performing an experiment or conducting a survey unless one also has one or more *alternate hypotheses* in mind.

PARAMETRIC, NONPARAMETRIC, AND SEMIPARAMETRIC MODELS

Models can be subdivided into two components, one systematic and one random. The systematic component can be a function of certain predetermined parameters (a parametric model), be parameter-free (nonparametric), or be a mixture of the two types (semiparametric). The definitions in the following section apply to the random component.

PARAMETRIC, NONPARAMETRIC, AND SEMIPARAMETRIC STATISTICAL PROCEDURES

Parametric statistical procedures concern the parameters of distributions of a known form. One may want to estimate the variance of a normal distribution or the number of degrees of freedom of a chisquare distribution. Student *t*, the *F* ratio, and maximum likelihood are typical parametric procedures.

Nonparametric procedures concern distributions whose form is unspecified. One might use a nonparametric procedure like the bootstrap to obtain an interval estimate for a mean or a median or to test that the distributions of observations drawn from two different populations are the same. Nonparametric procedures are often referred to as distribution-free, though not all distribution-free procedures are nonparametric in nature.

Semiparametric statistical procedures concern the parameters of distributions whose form is not specified. Permutation methods and *U* statistics are typically employed in a semiparametric context.

SIGNIFICANCE LEVEL AND *p* VALUE

The *significance level* is the probability of making a Type I error. It is a characteristic of a statistical procedure.

The *p value* is a random variable that depends both upon the sample and the statistical procedure that is used to analyze the sample.

If one repeatedly applies a statistical procedure at a specific significance level to distinct samples taken from the same population when the hypothesis is true and all assumptions are satisfied, then the *p* value will be less than or equal to the significance level with the frequency given by the significance level.

TYPE I AND TYPE II ERROR

A Type I error is the probability of rejecting the hypothesis when it is true. A Type II error is the probability of accepting the hypothesis when an alternative hypothesis is true. Thus, a Type II error depends on the alternative.

TYPE II ERROR AND POWER

The power of a test for a given alternative hypothesis is the probability of rejecting the original hypothesis when the alternative is true. A Type II error is made when the original hypothesis is accepted even though the alternative is true. Thus, power is one minus the probability of making a Type II error.

Bibliography

Adams DC; Gurevitch J; Rosenberg MS. Resampling tests for meta-analysis of ecological data. *Ecology* 1997; **78**:1277–1283.

Albers W; Bickel PJ; Van Zwet WR. Asymptotic expansions for the power of distribution-free tests in the one-sample problem. *Ann. Statist.* 1976; **4**:108–156.

Altman DG. Statistics in medical journals. *Stat. Med.* 1982; **1**:59–71.

Altman DG. Randomisation. *BMJ* 1991a; **302**:1481–1482.

Altman DG. Statistics in medical journals: Developments in the 1980s. *Stat. Med.* 1991b; **10**:1897–1913.

Altman DG. The scandal of poor medical research. *BMJ* 1994; **308**:283–284.

Altman DG. Statistical reviewing for medical journals. *Stat. Med.* 1998a; **17**: 2662–2674.

Altman DG. Commentary: Within trial variation—A false trail? *J. Clin. Epidemiol.* 1998b; **51**:301–303.

Altman DG. Statistics in medical journals: Some recent trends. *Stat. Med.* 2000; **19**:3275–3289.

Altman DG. Poor quality medical research: What can journals do? *JAMA* 2002; **287**:2765.

Altman DG; De Stavola BL; Love SB; Stepniewska KA. Review of survival analyses published in cancer journals. *Br. J. Cancer* 1995; **72**:511–518.

Altman DG; Lausen B; Sauerbrei W; Schumacher M. Dangers of using "optimal" cutpoints in the evaluation of prognostic factors [Commentary]. JNCI 1994; **86**:829–835.

Altman DG; Schulz KF; Moher D; Egger M; Davidoff F; Elbourne D; Gøtzsche PC; Lang T for the CONSORT Group. The revised consort statement for reporting randomized trials: Explanation and elaboration. *Ann. Int. Med.* 2001; **134**:663–694.

Aly E-E AA. Simple test for dispersive ordering. *Stat. Prob. Lett.* 1990; **9**:323–325.

Andersen B. *Methodological Errors in Medical Research.* Oxford: Blackwell, 1990.

Common Errors in Statistics (and How to Avoid Them), by Phillip I. Good and James W. Hardin.
ISBN 0-471-46068-0   Copyright © 2003 John Wiley & Sons, Inc.

Anderson DR; Burnham KP; Thompson WL. Null hypothesis testing: Problems, prevalence, and an alternative. *J. Wildlife Manage.* 2000; **64**:912–923.

Anderson S; Hauck WW. A proposal for interpreting and reporting negative studies. *Stat. Med.* 1986; **5**:203–209.

Anscombe F. Sequential medical trials [book review]. *JASA* 1963; **58**:365.

Armitage P. Test for linear trend in proportions and frequencies. *Biometrics* 1955; **11**:375–386.

Avram MJ; Shanks CA; Dykes MHM; Ronai AK; Stiers WM. Statistical methods in anesthesia articles: An evaluation of two American journals during two six-month periods. *Anesth. Analg.* 1985; **64**:607–611.

Bacchetti P. Peer review of statistics in medical research: The other problem. *BMJ* 2002; **324**:1271–1273.

Badrick TC; Flatman RJ. The inappropriate use of statistics, *NZ J Med Lab Sci.* 1999; **53**:95–103.

Bailar JC; Mosteller F. Guidelines for statistical reporting in articles for medical journals. Amplifications and explanations. *Ann. Intern. Med.* 1988; **108**:66–73.

Bailey KR. Inter-study differences: how should they influence the interpretation and analysis of results? *Stat. Med.* 1987; **6**:351–358.

Bailor AJ. Testing variance equality with randomization tests. *Stat. Comp. Simul.* 1989; **31**:1–8.

Baker RD. Two permutation tests of equality of variance. *Stat. Comput.* 1995; **5**(4):289–296.

Balakrishnan N; Ma CW. A comparative study of various tests for the equality of two population variances. *Stat. Comp. Simul.* 1990; **35**:41–89.

Barbui C; Violante A; Garattini S. Does placebo help establish equivalence in trials of new antidepressants? *Eur. Psychiatry* 2000; **15**:268–273.

Barnston AG; van den Dool HM. A degeneracy in cross-validated skill in regression-based forecasts. *J. Climate* 1993; **6**:963–977.

Barrodale I; Roberts FDK. An improved algorithm for discrete l_1 linear approximations. *Soc. Ind. Appl. Math. J. Numer. Anal.* 1973; **10**:839–848.

Bayarri MJ; Berger JQ. Quantifying surprise in the data and model verification. In: *Bayesian Statistics* 6. Bernado JM et al., eds. Oxford: Clarendon Press, 1999; 53–83 (with discussion).

Bayes T. An essay toward solving a problem in the doctrine of chances. *Philos. Trans. R. Soci.* 1763; **53**:370–418.

Begg CB; Cho M; Eastwood S; Horton R; Moher D; Olkin I; Pitkin R; Rennie D; Schulz KF; Simel D; Stroup DF. Improving the quality of reporting of randomized controlled trials: The CONSORT Statement. *JAMA* 1996; **276**:637–639.

Bent GC; Archfield SA. A logistic regression equation for estimating the probability of a stream flowing perennially in Massachusetts USGC. Water-Resources Investigations Report 02-4043, 2002.

Berger JO. *Statistical Decision Theory and Bayesian Analysis*, 2nd ed. New York: Springer-Verlag, 1986.

Berger JO. Could Fisher, Jefferies, and Neyman have agreed on testing? *Stat. Sci.* In press.

Berger JO; Berry DA. Statistical analysis and the illusion of objectivity. *Am. Sci.* 1988; **76**:159–165.

Berger JO; Sellke T. Testing a point null hypothesis: The irreconcilability of *P*-values and evidence. *JASA* 1987; **82**:112–122.

Berger V. Pros and cons of permutation tests. *Stat. Med.* 2000; **19**:1319–1328.

Berger VW. Improving the information content of endpoints in clinical trials. *Controlled Clin. Trials* 2002; **23**:502–514.

Berger VW, Exner DV. Detecting selection bias in randomized clinical trials. *Controlled Clin. Trials* 1999; **20**:319–327.

Berger VW; Ivanova A. Bias of linear rank tests for stochastic order in ordered categorical data. *J. Stat. Plann. Inference* 2002; **107**:1, in press.

Berger VW; Lunneborg C; Ernst MD; Levine JG. Parametric analyses in randomized clinical trials. *J. Mod. Appl. Stat. Methods* 2002; **1**:74–82.

Berger VW; Permutt T; Ivanova A. Convex hull test of ordered categorical data. *Biometrics* 1998; **54**:1541–1550.

Berkeley G. *Treatise Concerning the Principles of Human Knowledge.* Oxford: Oxford University Press, 1710.

Berkey C; Hoaglin D; Mosteller F; Colditz G. A random effects regression model for meta-analysis. *Stat. Med.* 1995; **14**:395–411.

Berkson J. Tests of significance considered as evidence. *JASA* 1942; **37**:325–335.

Berlin JA; Laird NM; Sacks HS; Chalmers TC. A comparison of statistical methods for combining event rates from clinical trials. *Stat. Med.* 1989; **8**:141–151.

Berry DA. Decision analysis and Bayesian methods in clinical trials. In: P Thall, ed. *Recent Advances in Clinical Trial Design and Analysis.* Kluwer Press, New York, 1995, pp. 125–154.

Berry DA. *Statistics: A Bayesian Perspective.* Belmont, CA: Duxbury Press, 1996.

Berry DA; Stangl DK. *Bayesian Biostatistics.* New York: Marcel Dekker, 1996.

Bickel P; Klassen CA; Ritov Y; Wellner J. *Efficient and Adaptive Estimation for Semiparametric Models.* Baltimore: Johns Hopkins University Press, 1993.

Bland JM; Altman DG. Comparing methods of measurement: Why plotting difference against standard method is misleading. *Lancet* 1995; **346**:1085–1087.

Block G. A review of validations of dietary assessment methods. *Am. J. Epidemiol.* 1982; **115**:492–505.

Bly RW. *The Copywriter's Handbook: A Step-By-Step Guide to Writing Copy That Sells.* New York: Henry Holt, 1990.

Bly RW. *Power-Packed Direct Mail: How to Get More Leads and Sales by Mail.* New York: Henry Holt, 1996.

Blyth CR. On the inference and decision models of statistics (with discussion). *Ann. Stat.* 1970; **41**:1034–1058.

Bothun G. *Modern Cosmological Observations and Problems.* London: Taylor and Francis, 1998.

Box GEP; Anderson SL. Permutation theory in the development of robust criteria and the study of departures from assumptions. *JRSS*-B 1955; **17**:1–34.

Box GEP; Tiao GC. A note on criterion robustness and inference robustness. *Biometrika* 1964; **51**:169–173.

Brockwell PJ; Davis RA. *Time Series: Theory and Methods*. New York: Springer-Verlag, 1987.

Browne MW. A comparison of single sample and cross-validation methods for estimating the mean squared error of prediction in multiple linear regression. *Br. J. Math. Stat. Psychol.* 1975; **28**:112–120.

Buchanan-Wollaston H. The philosophic basis of statistical analysis. *J. Int. Council Explor. Sea* 1935; **10**:249–263.

Cade B; Richards L. Permutation tests for least absolute deviation regression. *Biometrics* 1996; **52**:886–902.

Callaham ML; Wears RL; Weber EJ; Barton C; Young G. Positive-outcome bias and other limitations in the outcome of research abstracts submitted to a scientific meeting. *JAMA* 1998; **280**:254–257.

Camstra A; Boomsma A. Cross-validation in regression and covariance structure analysis. *Sociol. Methods Res.* 1992; **21**:89–115.

Canty AJ; Davison AC; Hinkley DV; Ventura V. Bootstrap diagnostics. http://www.stat.cmu.edu/www/cmu-stats/tr/tr726/tr726.html, 2000.

Capaldi D; Patterson GR. An approach to the problem of recruitment and retention rates for longitudinal research. *Behav. Assess.* 1987; **9**:169–177.

Cappuccio FP; Elliott P; Allender PS; Pryer J; Follman DA; Cutler JA. Epidemiologic association between dietary calcium intake and blood pressure: a meta-analysis of published data. *Am. J. Epidemiol.* 1995; **142**:935–945.

Carleton RA; Lasater TM; Assaf AR; Feldman HA; McKinlay S, et al. The Pawtucket Heart Health Program: Community changes in cardiovascular risk factors and projected disease risk. *Am. J. Public Health* 1995; **85**:777–785.

Carlin BP; Louis TA. *Bayes and Empirical Bayes Methods for Data Analysis*. London: Chapman and Hall, 1996.

Carmer SG; Walker WM. Baby bear's dilemma: A statistical tale. *Agron. J.* 1982; **74**:122–124.

Carpenter J; Bithell J. Bootstrap confidence intervals. *Stat. Med.* 2000; **19**:1141–1164.

Carroll RJ; Ruppert D. Transformations in regression: A robust analysis. *Technometrics* 1985; **27**:1–12.

Carroll RJ; Ruppert D. *Transformation and Weighting in Regression*. Boca Raton, FL: CRC, 2000.

Casella G; Berger RL. *Statistical Inference*. Pacific Grove, CA: Wadsworth & Brooks, 1990.

Chalmers TC. Problems induced by meta-analyses. *Stat. Med.* 1991; **10**:971–980.

Chalmers TC; Celano P; Sacks HS; Smith H. Bias in treatment assignment in controlled clinical trials. *New England J. Med.* 1983; **309**:1358–1361.

Chalmers TC; Frank CS; Reitman D. Minimizing the three stages of publication bias. *JAMA* 1990; **263**:1392–1395.

Charlton BG. The future of clinical research: From megatrials towards methodological rigour and representative sampling. *J. Eval. Clin. Pract.* 1996; **2**:159–169.

Cherry S. Statistical tests in publications of The Wildlife Society, *Wildlife Soc. Bull.* 1998; **26**:947–953.

Chiles JR. *Inviting Disaster: Lessons from the Edge of Technology.* New York: Harper-Collins, 2001.

Choi BCK. Development of indicators for occupational health and safety surveillance. *Asian-Pac. Newsl.* 2000; **7**.
http://www.occuphealth.fi/e/info/asian/ap100/develop04.htm.

Clemen RT. Combining forecasts: A review and annotated bibliography. *Int. J. Forecasting* 1989; **5**:559–583.

Clemen RT. *Making Hard Decisions.* Boston: PWS-Kent, 1991.

Clemen RT; Jones SK; Winkler RL. Aggregating forecasts: An empirical evaluation of some Bayesian methods. In: DA Berry, K Chaloner, eds. *Bayesian Analysis in Statistics and Econometrics.* New York: John Wiley & Sons, 1996, pp. 3–13.

Cleveland WS. The *Elements of Graphing Data.* Summit, NJ: Hobart Press, 1985.

Cleveland WS; McGill ME. *Dynamic Graphics Statistics.* London: CRC Press, 1988.

Cochran WG. *Sampling Techniques*, 3rd ed. New York: John Wiley & Sons, 1977.

Cohen J. Things I have learned (so far). *Am. Psychol.* 1990; **45**:1304–1312.

Collins R; Keech A; Peto R; Sleight P; Kjekshus J; Wilhelmsen L; et al. Cholesterol and total mortality: Need for larger trials. *BMJ* 1992; **304**:1689.

Conover W; Salsburg D. *Biometrics* 1988; **44**:189–196.

Conover WJ; Johnson ME; Johnson MM. Comparative study of tests for homogeneity of variances: With applications to the outer continental shelf bidding data. *Technometrics* 1981; **23**:351–361.

Converse JM; Presser S. *Survey Questions: Handcrafting the Standardized Questionaire.* London: Sage, 1986.

Cooper HM; Rosenthal R. Statistical versus traditional procedures for summarising research findings. *Psychol. Bull.* 1980; **87**:442–449.

Copas JB; Li HG. Inference for non-random samples (with discussion). *JRSS* 1997; **59**:55–95.

Cornfield J; Tukey JW. Average values of mean squares in factorials. *Ann. Math. Stat.* 1956; **27**:907–949.

Cox DR. Some problems connected with statistical inference. *Ann. Math. Stat.* 1958; **29**:357–372.

Cox DR. The role of significance tests. *Scand. J. Stat.* 1977; **4**:49–70.

Cox DR. Some remarks on consulting. *Liaison* (Statistical Society of Canada) 1999; **13**:28–30.

Cummings P; Koepsell TD. Statistical and design issues in studies of groups. *Inj. Prev.* 2002; **8**:6–7.

Dar R; Serlin; Omer H. Misuse of statistical tests in three decades of psychotherapy research. *J. Consult. Clin. Psychol.* 1994; **62**:75–82.

Davison AC; Hinkley DV. *Bootstrap Methods and Their Application.* New York: Cambridge University Press, 1997.

Davison AC; Snell EJ. Residuals and diagnostics. In: DV Hinkley, N Reid, and EJ Shell, eds. *Statistical Theory and Modelling,* London: Chapman and Hall, 1991, p. 83.

Day S. Blinding or masking. In: P. Armitage and T. Colton, eds. *Encyclopedia of Biostatistics,* Vol. 1, Chichester: John Wiley & Sons, 1998.

DeGroot MH. *Optimal Statistical Decisions.* New York: McGraw-Hill, 1970.

Delucchi KL. The use and misuse of chisquare. Lewis and Burke revisited. *Psych. Bull.* 1983; **94**:166–176.

Diaconis P. Statistical problems in ESP research. *Science* 1978; **201**:131–136.

Diciccio TJ; Romano JP. A review of bootstrap confidence intervals (with discussion). *JRSS B* 1988; **50**:338–354.

Dixon PM. Assessing effect and no effect with equivalence tests. In: MC Newman and CL Strojan, eds. *Risk Assessment: Logic and Measurement.* Chelsea, MI: Ann Arbor Press, 1998.

Djulbegovic B; Lacevic M; Cantor A; Fields KK; Bennett CL; Adams JR; Kuderer NM; Lyman GH. The uncertainty principle and industry-sponsored research. *Lancet* 2000; **356**:635–638.

Donner A; Brown KS; Brasher P. A methodological review of non-therapeutic intervention trials employing cluster randomization, 1979–1989. *Int J. Epidemiol.* 1990; **19**:795–800.

Duggan TJ; Dean CW. Common misinterpretations of significance levels in sociological Journals. *Am. Sociol.* 1968; **February**:45–46.

Dyke G. How to avoid bad statistics. *Field Crops Res.* 1997; **51**:165–197.

Easterbrook PJ; Berlin JA; Gopalan R, Matthews DR. Publication bias in clinical research. *Lancet* 1991; **337**:867–872.

Ederer F. Why do we need controls? Why do we need to randomize? *Amer. J. Ophthl.* 1975; **76**:758–762.

Edwards W; Lindman H; Savage L. Bayesian statistical inference for psychological research. *Psychol Rev.* 1963; **70**:193–242.

Efron B. Bootstrap methods, another look at the jackknife. *Annals Stat.* 1979; 7:1–26.

Efron B. *The Jackknife, the Bootstrap, and Other Resampling Plans.* Philadelphia: SIAM, 1982.

Efron B. Better bootstrap confidence intervals, (with discussion). *JASA* 1987; **82**:171–200.

Efron B. Bootstrap confidence intervals: Good or bad? (with discussion). *Psychol. Bull.* 1988; **104**:293–296.

Efron B. Six questions raised by the bootstrap. In: R LePage and L Billard, eds. *Exploring the Limits of the Bootstrap.* New York: John Wiley & Sons, 1992.

Efron B; Morris C. Stein's paradox in statistics. *Sci. Am.* 1977; **236**:119–127.

Efron B; Tibshirani R. Bootstrap methods for standard errors, confidence intervals, and other measures of statistical accuracy. *Stat. Sci.* 1986; **1**:54–77.

Efron B; Tibshirani R. *An Introduction to the Bootstrap.* New York: Chapman and Hall, 1993.

Egger M; Smith GD. Meta-analysis: Potentials and promise. *BMJ* 1997; **315**: 1371–1374.

Egger M; Schneider M; Smith GD. Spurious precision? Meta-analysis of observational studies. *Br. Med. J.* 1998; **316**:140–143.

Egger M; Smith GD; Phillips AN. Meta-analysis: Principles and procedures. *BMJ* 1997; **315**:1533–1537.

Ehrenberg ASC. Rudiments of numeracy. *JRSS Ser. A* 1977; **140**:277–297.

Elwood JM. *Critical Appraisal of Epidemiological Studies and Clinical Trials*, 2nd ed. New York: Oxford University Press, 1998.

Eysenbach G; Sa E-R. Code of conduct is needed for publishing raw data. *BMJ* 2001; **323**:166.

Fears TR; Tarone RE; Chu KC. False-positive and false-negative rates for carcinogenicity screens. *Cancer Res.* 1977; **37**:1941–1945.

Feinstein AR. P-values and confidence intervals: Two sides of the same unsatisfactory coin. *J. Clin. Epidemiol.* 1998; **51**:355–360.

Feinstein AR; Concato J. The quest for "power": contradictory hypotheses and inflated sample sizes. *J. Clin. Epidemiol.* 1998; **51**:537–545.

Feller W. *An Introduction to Probability Theory and Its Applications*. Vol. 2. New York: John Wiley & Sons, 1966.

Felson DT; Anderson JJ; Meenan RF. The comparative efficacy and toxicity of second-line drugs in rheumatoid arthritis. *Arthritis Rheum.* 1990; **33**:1449–1461.

Felson DT; Cupples LA; Meenan RF. Misuse of statistical methods in *Arthritis and Rheumatism.* 1982 versus 1967–68. *Arthritis and Rheum.* 1984; **27**: 1018–1022.

Feng Z; Grizzle J. Correlated binomial variates: properties of estimator of ICC and its effect on sample size calculation. *Stat. Med.* 1992; **11**:1607–1614.

Feng Z; McLerran D; Grizzle J. A comparison of statistical methods for clustered data analysis with Gaussian error. *Stat. Med.* 1996; **15**:1793–1806.

Feng Z; Diehr P; Peterson A; McLerran D. Selected statistical issues in group randomized trials. *Annu. Rev. Public Health* 2001; **22**:167–187.

Fienberg SE. Damned lies and statistics: misrepresentations of honest data. In: Editorial Policy Committee. *Ethics and Policy in Scientific Publications.* Reston, VA: Council of Biology Editors, 1990, pp. 202–206.

Fink A; Kosecoff JB. *How to Conduct Surveys: A Step by Step Guide.* London: Sage, 1988.

Finney DJ. The responsible referee. *Biometrics* 1997; **53**:715–719.

Firth D. General linear models. In: DV Hinkley, N Reid, and EJ Shell, eds. *Statistical Theory and Modelling,* London: Chapman and Hall, 1991, p. 55.

Fisher NI; Hall P. On bootstrap hypothesis testing. *Aust. J. Stat.* 1990; **32**:177–190.

Fisher NI; Hall P. Bootstrap algorithms for small samples. *J. Stat. Plan Infer.* 1991; **27**:157–169.

Fisher RA. *Statistical Methods for Research Workers.* Edinburgh: Oliver & Boyd; 1st ed 1925.

Fisher RA. *Design of Experiments*. New York: Hafner, 1935.

Fisher RA. *Statistical Methods and Scientific Inference*, 3rd ed. New York: Macmillan, 1973.

Fleming TR. Surrogate markers in AIDs and cancer trials. *Stat. Med.* 1995; **13**: 1423–1435.

Fligner MA; Killeen TJ. Distribution-free two-sample tests for scale. *JASA* 1976; **71**:210–212.

Fowler FJ Jr; Fowler FJ. *Improving Survey Questions: Design and Evaluation*. London: Sage, 1995.

Frank D; Trzos RJ; Good P. Evaluating drug-induced chromosome alterations. *Mutat. Res.* 1978; **56**:311–317.

Freedman DA. A note on screening regression equations. *Am. Stat.* 1983; **37**: 152–155.

Freedman DA. From association to causation. *Stat. Sci.* 1999; **14**:243–258.

Freedman DA; Navidi W; Peters SC. On the impact of variable selection in fitting regression equations. In: TK Dijkstra, ed. *On Model Uncertainty and Its Statistical Implications*. Berlin: Springer, 1988, pp. 1–16.

Friedman LM; Furberg CD; DeMets DL. *Fundamentals of Clinical Trials*, 3rd ed. St. Louis: Mosby, 1996.

Friedman M. The use of ranks to avoid the assumption of normality implicit in the analysis of variance. *JASA* 1937; **32**:675–701.

Freiman JA; Chalmers TC; Smith H; Kuebler RR. The importance of beta; the type II error; and sample size in the design and interpretation of the randomized controlled trial. In: JC Bailar and F Mosteller, eds. *Medical Uses of Statistics*. Boston: NEJM Books, 1992, p. 357.

Fritts HC; Guiot J; Gordon GA. Verification. In: ER Cook and LA Kairiukstis, eds. *Methods of Dendrochronology; Applications in the Environmental Sciences*. Norwell, MA: Kluwer Academic Publishers, 1990, pp. 178–185.

Gail MH; Byar DP; Pechacek TF; Corle DK. Aspects of statistical design for the Community Intervention Trial for Smoking Cessation (COMMIT). *Controlled Clin. Trials* 1992; **123**:6–21.

Gail MH; Mark SD; Carroll R; Green S; Pee D. On design considerations and randomization-based inference for community intervention trials. *Stat. Med.* 1996; **15**:1069–1092.

Gail MH; Tan WY; Piantadosi S. Tests for no treatment effect in randomized clinical trials. *Biometrika*. 1988; **75**:57–64.

Gardner, MJ; Altman DG. Confidence intervals rather than P values: Estimation rather than hypothesis testing. *BMJ* 1996; **292**:746–750.

Gardner MJ; Bond J. An exploratory study of statistical assessment of papers published in the *Journal of the American Medical Association*. *JAMA* 1990; **263**: 1355–1357.

Gardner MJ; Machin D; Campbell MJ. Use of check lists in assessing the statistical content of medical studies. *BMJ* 1986; **292**:810–812.

Garthwaite PH. Confidence intervals from randomization tests. *Biometrics* 1996; **52**:1387–1393.

Gastwirth JL; Rubin H. Effect of dependence on the level of some one-sample tests. *JASA* 1971; **66**:816–820.

Gavarret J. *Principes Généraux de Statistique Médicale.* Paris: Libraires de la Faculte de Medecine de Paris, 1840.

Geary RC. Testing normality. *Biometrika* 1947; **34**:241.

George SL. Statistics in medical journals: a survey of current policies and proposals for editors. *Med. Pediatr. Oncol.* 1985; **13**:109–112.

Geweke JK; DeGroot MH. *Optimal Statistical Decisions.* New York: McGraw-Hill, 1970.

Gigerenzer G. *Calculated Risks: How To Know When Numbers Deceive You.* New York: Simon & Schuster, 2002.

Gill J. Whose variance is it anyway? Interpreting empirical models with state-level data. *State Polit. Policy Q.* 2001; **Fall**:318–338.

Gillett R. Meta-analysis and bias in research reviews. *J. Reprod. Infant Psychol.* 2001; **19**:287–294.

Gine E; Zinn J. Necessary conditions for a bootstrap of the mean. *Ann. Stat.* 1989; **17**:684–691.

Glantz S. Biostatistics: How to detect, correct and prevent errors in the medical literature. *Circulation* 1980; **61**:1–7.

Glass GV; Peckham PD; Sanders JR. Consequences of failure to meet the assumptions underlying the fixed effects analysis of variance and covariance. *Rev. Educ. Res.* 1972; **42**:237–288.

Goldberger, AS. Note on stepwise least squares. *JASA* 1961; **56**:105–110.

Gong G. Cross-validation, the jackknife, and the bootstrap: Excess error in forward logistic regression. *JASA* 1986; **81**:108–113.

Good IJ. *Probability and the Weighing of Evidence.* London: Griffin, 1950.

Good IJ. The Bayes/non-Bayes compromise: A brief review. *JASA* 1992; **87**:597–606.

Good PI. *Permutation Tests*, 2nd ed. New York: Springer, 2000.

Good PI. *Resampling Methods*, 2nd ed. Boston: Birkhauser, 2001a.

Good PI. *Applications of Statistics in the Courtroom.* London: Chapman and Hall, 2001b.

Good PI. Extensions of the concept of exchangeability and their applications to testing hypotheses. *J. Mod. Appl. Statist. Meth.* 2002; **2**:243–247

Goodman SN. Towards evidence-based medical statistics. II. The Bayes Factor. *Ann. Intern. Med.* 1999; **130**:1005–1013.

Goodman SN. Of *p*-values and Bayes: A modest proposal. *Epidemiology* 2001; **12**:295–297.

Goodman SN; Altman DG; George SL. Statistical reviewing policies of medical journals: Caveat lector? *J. Gen. Intern. Med.* 1998; **13**:753–756.

Gore S; Jones IG; Rytter EC. Misuse of statistical methods: critical assessment of articles in *BMJ* from January to March 1976. *BMJ* 1977; **1**:85–87.

Götzsche PC. Reference bias in reports of drug trials. *BMJ* 1987; **295**:654–656.

Götzsche PC; Podenphant J; Olesen M; Halberg P. Meta-analysis of second-line antirheumatic drugs: Sample size bias and uncertain benefit. *J. Clin. Epidemiol.* 1992; **45**:587–594.

Grant A. Reporting controlled trials. *Br. J. Obstet. Gynaecol.* 1989; **96**:397–400.

Graumlich L. A 1000-year record of temperature and precipitation in the Sierra Nevada, *Quaternary Res.* 1993; **39**:249–255.

Green PJ; Silverman BW. *Nonparametric Regression and Generalized Linear Models.* London: Chapman and Hall, 1994.

Greenland S. Modeling and variable selection in epidemiologic analysis. *Am. J. Public Health* 1989; **79**:340–349.

Greenland S. Randomization, statistics, and causal inference. *Epidemiology* 1990; **1**:421–429.

Greenland S. Probability logic and probabilistic induction [see comments]. *Epidemiology* 1998; **9**:322–332.

Gurevitch J; Hedges LV. Meta-analysis: Combining the results of independent studies in experimental ecology. In: S Scheiner and J Gurevitch, eds. *The Design and Analysis of Ecological Experiments.* London: Chapman & Hall, 1993, pp. 378–398.

Guthery FS; Lusk JJ; Peterson MJ. The fall of the null hypothesis: Liabilities and opportunities. *J. Wildlife Manage.* 2001; **65**:379–384.

Guttorp P. *Stochastic Modeling of Scientific Data.* London: Chapman & Hall, 1995.

Hagood MJ. *Statistics for Sociologists.* New York: Reynal and Hitchcock, 1941.

Hall P; Wilson SR. Two guidelines for bootstrap hypothesis testing. *Biometrics* 1991; **47**:757–762.

Hardin JW; Hilbe JM. *Generalized Estimating Equations.* London: Chapman and Hall/CRC, 2003.

Harley SJ; Myers RA. Hierarchical Bayesian models of length-specific catchability of research trawl surveys. *Can. J. Fish. Aquat. Sci.* 2001; **58**:1569–1584.

Harrell FE; Lee KL. A comparison of the discrimination of discriminant analysis and logistic regression under multivariate normality. In: PK Sen, ed. *Biostatistics: Statistics in Biomedical; Public Health; and Environmental Sciences. The Bernard G. Greenberg Volume.* New York: North-Holland, 1985, pp. 333–343.

Harrell FE; Lee KL; Mark DB. Multivariable prognostic models: Issues in developing models; evaluating assumptions and adequacy; and measuring and reducing errors. *Stat. Med.* 1996; **15**:361–387.

Hastie T; Tibshirani R; Friedman JH. *The Elements of Statistical Learning : Data Mining, Inference, and Prediction.* New York: Springer, 2001.

Hedges LV; Olkin I. *Statistical Methods for Meta-Analysis.* New York: Academic Press, 1985.

Hertwig R; Todd PM. Biases to the left, fallacies to the right: Stuck in the middle with null hypothesis significance testing (with discussion). *Psycoloquy* 2000; **11**:28.

Hilton J. The appropriateness of the Wilcoxon test in ordinal data. *Stat. Med.* 1996; **15**:631–645.

Hinkley DV; Shi S. Importance sampling and the nested bootstrap. *Biometrika*. 1989; **76**:435–446.

Hodges JS. Uncertainty, policy analysis and statistics. *Statist. Sci.* 1987; **2**:259–291.

Hoenig JM; Heisey DM. The abuse of power: The pervasive fallacy of power calculations for data analysis. *Am. Stat.* 2001; **55**:19–24.

Horowitz RI. Large scale randomised evidence; large simple trials and overviews of trials: Discussion—A clinician's perspective on meta-analysis. *J. Clin. Epidemiol.* 1995; **48**:41–44.

Horwitz RI; Singer BH; Makuch RW; Viscolia CM. Clinical versus statistical considerations in the design and analysis of clinical research. *J. Clin. Epidemiol.* 1998; **51**:305–307.

Hosmer DW; Lemeshow SL. *Applied Logistic Regression*. New York: John Wiley & Sons, 2001.

Huber PJ. *Robust Statistics*. New York: John Wiley & Sons, 1981.

Hume D. *An Enquiry Concerning Human Understanding*. Oxford: Oxford University Press, 1748.

Hungerford TW. *Algebra*. New York: Holt, Rinehart, and Winston, 1974.

Hunter JE; Schmidt FL. Eight common but false objections to the discontinuation of significance testing in the analysis of research data. In: LL Harlow, SA Mulaik and JH Steiger, eds. *What If There Were No Significance Tests?* Mahwah, NJ: Lawrence Erlbaum Associates, 1997, pp. 37–64.

Hurlbert SH. Pseudoreplication and the design of ecological field experiments. *Ecol. Monogr.* 1984; **54**:198–211.

Husted JA; Cook RJ; Farewell VT; Gladman DD. Methods for assessing responsiveness: A critical review and recommendations. *J. Clin. Epidemiol.* 2000; **53**: 459–468.

Hutchon DJR. Infopoints: Publishing raw data and real time statistical analysis on e-journals. *BMJ* 2001; **322**:530.

International Committee of Medical Journal Editors. Uniform requirements for manuscripts submitted to biomedical journals. *JAMA* 1997; **277**:927–934.

International Study of Infarct Survival Collaborative Group. Randomized trial of intravenous streptokinase, oral aspirin, both or neither, among 17187 cases of suspected acute myocardial infarction. ISIS-2. *Lancet* 1988; **2**:349–362.

Jennison C; Turnbull BW. *Group Sequential Methods with Applications to Clinical Trials*. Boca Raton, FL: CRC, 1999.

Johnson DH. The insignificance of statistical significance testing. *J. Wildlife Manage.* 1999; **63**:763–772.

Jones LV. Statistics and research design. *Annu. Rev. Psychol.* 1955; **6**:405–430.

Jones LV; Tukey JW. A sensible formulation of the significance test. *Psychol. Methods* 2000; **5**:411–416.

Kadane IB; Dickey J; Winklcr R; Smith W; Peters S. Interactive elicitation of opinion for a normal linear model. *JASA* 1980; **75**:845–854.

Kanarek MS; Conforti PM; Jackson LA; Cooper RC; Murchio JC. Asbestos in drinking water and cancer incidence in the San Francisco Bay Area. *Am. J. Epidemiol.* 1980; **112**:54–72.

Kaplan J. Misuses of statistics in the study of intelligence: the case of Arthur Jensen (with disc). *Chance* 2001; **14**:14–26.

Kass R; Raftery A. Bayes factors. *JASA* 1995; **90**:773–795.

Kaye DH. *Plemmel* as a primer on proving paternity. *Willamette L. Rev.* 1988; **24**:867.

Keynes JM. *A Treatise on Probability.* London: Macmillan, 1921.

Knight K. On the bootstrap of the sample mean in the infinite variance case. *Annal Stat.* 1989; **17**:1168–1173.

Lachin JM. Sample size determination. In: P Armitage and T Colton, eds. *Encyclopedia of Biostatistics*, 5. Chichester: John Wiley & Sons, 1998, pp. 3892–3903.

Lambert D. Robust two-sample permutation tests. *Ann. Stat.* 1985; **13**:606–625.

Lang TA; Secic M. *How to Report Statistics in Medicine.* Philadelphia: American College of Physicians, 1997.

Lehmann EL. *Testing Statistical Hypotheses*, 2nd ed. New York: John Wiley & Sons, 1986, pp. 203–213 on robustness.

Lehmann EL. The Fisher, Neyman–Pearson theories of testing hypotheses: one theory or two? *JASA* 1993; **88**:1242–1249.

Lehmann EL; Casella G. *Theory of Point Estimation*, 2nd ed. New York: Springer, 1998.

Lehmann EL; D'Abrera HJM. *Nonparametrics: Statistical Methods Based on Ranks*, 2nd ed. New York: McGraw-Hill, 1988.

Leizorovicz A; Haugh MC; Chapuis F-R; Samama MM; Boissel J-P. Low molecular weight heparin in prevention of perioperative thrombosis. *BMJ* 1992; **305**:913–920.

Lettenmaier DP. Space-time correlation and its effect on methods for detecting aquatic ecological change. *Can. J. Fish. Aquat. Sci.* 1985; **42**:1391–1400. Correction. 1986; **43**:1680.

Lewis D; Burke CJ. Use and misuse of the chi-square test. *Psychol. Bull.* 1949; **46**:433–489.

Lieberson S. *Making It Count.* Berkeley: University of California Press, 1985.

Light RJ; Pillemer DB. *Summing Up: The Science of Reviewing Research.* Cambridge; MA: Harvard University Press, 1984.

Lindley DV. The choice of sample size. *The Statistician* 1997; **46**:129–138, 163–166.

Lindley D. The philosophy of statistics (with discussion). *The Statistician* 2000; **49**:293–337.

Lissitz RW; Chardos S. A study of the effect of the violation of the assumption of independent sampling upon the type I error rate of the two group *t*-test. *Educ. Psychol. Meas.* 1975; **35**:353–359.

Little RJA; Rubin DB. *Statistical Analysis with Missing Data.* New York: John Wiley & Sons, 1987.

Loader C. *Local Regression and Likelihood.* New York: Springer, 1999.

Locke J. *Essay Concerning Human Understanding*, 4th ed. Essex, England: Prometheus Books, 1700.

Lonergan JF. *Insight: A Study of Human Understanding*. University of Toronto Press, 1992.

Lord FM. Statistical adjustment when comparing preexisting groups. *Psychol. Bull.* 1969; **72**:336–337.

Lovell DJ; Giannini EH; Reiff A; Cawkwell GD; Silverman ED; Nocton JJ; Stein LD; Gedalia A; Ilowite NT; Wallace CA; Whitmore J; Finck BK: The Pediatric Reumatology Collaborative Study Group. Etanercept in children with polyarticular juvenile rheumatoid arthritis. *N. Engl. J. Med.* 2000; **342**:763–769.

MacArthur RD; Jackson GG. An evaluation of the use of statistical methodology in the *Journal of Infectious Diseases*. *J. Infect. Dis.* 1984; **149**:349–354.

Mangel M; Samaniego FJ. Abraham Wald's work on aircraft survivability. *JASA* 1984; **79**:259–267.

Manly BFJ. *Randomization, Bootstrap and Monte Carlo Methods in Biology*, 2nd ed. London: Chapman & Hall, 1997.

Manly B; Francis C. Analysis of variance by randomization when variances are unequal. *Aust. N. Z. J. Stat.* 1999; **41**:411–430.

Maritz JS. *Distribution Free Statistical Methods*, 2nd ed. London: Chapman & Hall, 1996.

Matthews JNS, Altman DG. Interaction 2: Compare effect sizes not p values. *BMJ* 1996; **313**:808.

Mayo DG. *Error and the Growth of Experimental Knowledge*. Chicago: University of Chicago Press, 1996.

McBride GB; Loftis JC; Adkins NC. What do significance tests really tell us about the environment? *Environ. Manage.* 1993; **17**:423–432. (erratum **19**:317).

McGuigan SM. The use of statistics in the *British Journal of Psychiatry*. *Br. J. Psychiatry* 1995; **167**:683–688.

McKinney PW; Young MJ; Hartz A; Bi-Fong Lee M. The inexact use of Fisher's exact test in six major medical journals. *JAMA* 1989; **261**:3430–3433.

Mena EA; Kossovsky N; Chu C; Hu C. Inflammatory intermediates produced by tissues encasing silicone breast prostheses. *J. Invest Surg.* 1995; **8**:31–42.

Michaelsen J. Cross-validation in statistical climate forecast models. *J. Climate Appl. Meteorol.* 1987; **26**:1589–1600.

Mielke PW; Berry KJ. *Permutation Methods: A Distance Function Approach*. New York: Springer, 2001.

Mielke PW; KJ Berry. Permutation covariate analyses of residuals based on Euclidean distance. *Psychol. Rep.* 1997; **81**:795–802.

Mielke PW; Berry KJ; Landsea CW; Gray WM. Artificial skill and validation in meteorological forecasting. *Weather and Forecasting* 1996; **11**:153–169.

Mielke PW; Berry KJ; Landsea CW; Gray WM. A single sample estimate of shrinkage in meteorological forecasting. *Weather and Forecasting* 1997; **12**:847–858.

Miller ME; Hui SL; Tierney WM. Validation techniques for logistic regression models. *Stat. Med.* 1991; **10**:1213–1226.

Miller RG. Jackknifing variances. *Ann. Math. Stat.* 1968; **39**:567–582.

Miller RG. *Beyond Anova: Basics of Applied Statistics*. New York: John Wiley & Sons, 1986.

Moher D; Cook DJ; Eastwood S; Olkin I; Rennie D; Stroup D for the QUOROM Group. Improving the quality of reports of meta-analyses of randomised controlled trials: The QUOROM statement. *Lancet* 1999; **354**:1896–1900.

Moiser CI. Symposium: The need and means of cross-validation, I: problems and design of cross-validation. *Educ. Psych. Measure.* 1951; **11**:5–11.

Montgomery DC; Myers RH. *Response Surface Methodology: Process and Product Optimization Using Designed Experiments.* New York: John Wiley & Sons, 1995.

Morris RW. A statistical study of papers in the *J. Bone and Joint Surgery BR. J. Bone Joint Surg. BR* 1988; **70B**:242–246.

Morrison DE; Henkel RE. *The Significance Test Controversy.* Chicago: Aldine, 1970.

Mosteller F. Problems of omission in communications. *Clin. Pharmacol. Ther.* 1979; **25**:761–764.

Mosteller F; Chalmers TC. Some progress and problems in meta-analysis of clinical trials. *Stat. Sci.* 1992; **7**:227–236.

Mosteller F; Tukey JW. *Data Analysis and Regression: A Second Course in Statistics.* Menlo Park: CA: Addison-Wesley, 1977.

Mulrow CD. The medical review article: state of the science. *Ann. Intern. Med.* 1987; **106**:485–488.

Murray GD. Statistical guidelines for the *British Journal of Surgery. Br. J. Surg.* 1991; **78**:782–784.

Murray GD. The task of a statistical referee. *Br. J. Surg.* 1988; **75**:664–667.

Nelder JA; Wedderburn RWM. Generalized linear models. *JRSS A* 1972; **135**:370–384.

Nester M. An applied statistician's creed. *Appl. Stat.* 1996; **45**:401–410.

Neyman J. *Lectures and Conferences on Mathematical Statistics and Probability,* 2nd ed., Washington, DC: Graduate School, U.S. Department of Agriculture, 1952.

Neyman J. Silver jubilee of my dispute with Fisher. *J. Oper. Res. Soc. Jpn.* 1961; **3**:145–154.

Neyman J. Frequentist probability and frequentist statistics. *Synthese* 1977; **36**:97–131.

Neyman J; Pearson ES. On the testing of specific hypotheses in relation to probability a priori. *Proc. Cambridge Philos. Soc.* 1933; **29**:492–510.

Neyman J; Pearson ES. On the problem of the most efficient tests of statistical hypotheses. *Philos. Trans. R. Soc. A.* 1933; **231**:289–337.

Nurmohamed MT; Rosendaal FR; Bueller HR; Dekker E; Hommes DW; Vandenbroucke JP; et al. Low-molecular-weight heparin versus standard heparin in general and orthopaedic surgery: A meta-analysis. *Lancet* 1992; **340**:152–156.

O'Brien PC. The appropriateness of analysis of variance and multiple-comparison procedures. *Biometrics* 1983; **39**:787–788.

O'Brien P. Comparing two samples: extension of the t, rank-sum, and log-rank tests. *JASA* 1988; **83**:52–61.

Oldham PD. A note on the analysis of repeated measurements of the same subjects. *J. Chron. Dis.* 1962; **15**:969–977.

Osborne J; Waters E. Four assumptions of multiple regression that researchers should always test. *Pract. Assess. Res. Eval.* 2002; **8**(2). Available online.

Padaki PM. Inconsistencies in the use of statistics in horticultural research. *Hort. Sci.* 1989; **24**:415.

Palmer RF; Graham JW; White EL; Hansen WB. Applying multilevel analytic strategies in adolescent substance use prevention research. *Prevent. Med.* 1998; **27**:328–336.

Pankratz A. *Forecasting with Dynamic Regression Models.* New York: John Wiley & Sons, 1991.

Parkhurst DF. Arithmetic versus geometric means for environmental concentration data. *Environmental Sci. Technol.* 1998; **32**:92A–98A.

Parkhurst DF. Statistical significance tests: Equivalence and reverse tests should reduce misinterpretation. *Bioscience* 2001; **51**:1051–1057.

Pesarin F. *Multivariate Permutation Tests.* New York: John Wiley & Sons, 2001.

Pettitt AN; Siskind V. Effect of within-sample dependence on the Mann–Whitney–Wilcoxon statistic. *Biometrika* 1981; **68**:437–441.

Phipps MC. Small samples and the tilted bootstrap. *Theory of Stochastic Processes* 1997; **19**:355–362.

Picard RR; Berk KN. Data splitting. *Am. Stat.* 1990; **44**:140–147.

Picard RR; Cook RD. Cross-validation of regression models. *JASA* 1984; **79**:575–583.

Pierce CS. *Values in a University of Chance*, PF Wiener, ed. New York: Doubleday Anchor Books, 1958.

Pilz J. *Bayesian Estimation and Experimental Design In Linear Regression Models*, 2nd ed. New York: John Wiley & Sons, 1991.

Pinelis IF. On minimax risk. *Theory Prob. Appl.* 1988; **33**:104–109.

Pitman EJG. Significance tests which may be applied to samples from any population. *R. Stat. Soc. Suppl.* 1937; **4**:119–130, 225–232.

Pitman EJG. Significance tests which may be applied to samples from any population. Part III. The analysis of variance test. *Biometrika* 1938; **29**:322–335.

Poole C. Beyond the confidence interval. *Am. J. Public Health* 1987; **77**:195–199.

Poole C. Low *p*-values or narrow confidence intervals: Which are more durable? *Epidemiology* 2001; **12**:291–294.

Praetz P. A note on the effect of autocorrelation on multiple regression statistics. *Aust. J. Stat.* 1981; **23**:309–313.

Proschan MA; Waclawiw MA. Practical guidelines for multiplicity adjustment in clinical trials. *Controlled Clin. Trials* 2000; **21**:527–539.

Ravnskov U. Cholesterol lowering trials in coronary heart disease: Frequency of citation and outcome. *BMJ* 1992; **305**:15–19.

Rea LM; Parker RA; Shrader A. *Designing and Conducting Survey Research: A Comprehensive Guide*, 2nd ed. Jossey-Bass, 1997.

Redmayne M. Bayesianism and proof. In: M Freeman and H Reece, eds., *Science in Court*, Brookfield, MA: Ashgate, 1998.

Reichenbach H. *The Theory of Probability*. Berkeley: University of California Press, 1949.

Rencher AC; Pun F-C. Inflation of R^2 in best subset regression. *Technometrics* 1980; **22**:49–53.

Rosenbaum PR. *Observational Studies*, 2nd ed. New York: Springer, 2002.

Rosenberger W; Lachin JM. *Randomization in Clinical Trials: Theory and Practice*. New York: John Wiley & Sons, 2002.

Rothman KJ. Epidemiologic methods in clinical trials. *Cancer* 1977; **39**:1771–1775.

Rothman KJ. No adjustments are needed for multiple comparisons. *Epidemiology* 1990a; **1**:43–46.

Rothman KJ. Statistics in nonrandomized studies. *Epidemiology* 1990b; **1**:417–418.

Roy J. Step-down procedure in multivariate analysis. *Ann. Math. Stat.* 1958; **29**(4):1177–1187.

Royall RM. *Statistical Evidence: A Likelihood Paradigm*. New York: Chapman and Hall, 1997.

Rozeboom W. The fallacy of the null hypothesis significance test. *Psychol. Bull.* 1960; **57**:416–428.

Salmaso L. Synchronized permutation tests in $2k$ factorial designs. *Commun. Stats. Theory Meth.* 2003 (in Press).

Savage LJ. *The Foundations of Statistics*. New York: Dover Publications, 1972.

Schmidt FL. Statistical significance testing and cumulative knowledge in psychology: Implications for training of researchers. *Psychol. Methods* 1996; **1**:115–129.

Schenker N. Qualms about bootstrap confidence intervals. *JASA* 1985; **80**:360–361.

Schor S; Karten I. Statistical evaluation of medical manuscripts. *JASA* 1966; **195**: 1123–1128.

Schroeder YC. The procedural and ethical ramifications of pretesting survey questions. *Am. J. Trial Advocacy* 1987; **11**:195–201.

Schulz KF. Subverting randomization in controlled trials. *JAMA* 1995; **274**:1456–1458.

Schulz KF. Randomised trials, human nature, and reporting guidelines. *Lancet* 1996; **348**:596–598.

Schulz KF; Chalmers I; Hayes R; Altman DG. Empirical evidence of bias. Dimensions of methodological quality associated with estimates of treatment effects in controlled trials. *JAMA* 1995; **273**:408–412.

Schulz KF, Grimes DA. Blinding in randomized trials: Hiding who got what. *Lancet* 2002; **359**:696–700.

Seidenfeld T. *Philosophical Problems of Statistical Inference*. Boston: Reidel, 1979.

Selike T; Bayarri MJ; Berger JO. Calibration of p-values for testing precise null hypotheses. *Am. Stat.* 2001; **55**:62–71.

Selvin H. A critique of tests of significance in survey research. *Am. Soc. Rev.* 1957; **22**:519–527.

Senn S. A personal view of some controversies in allocating treatment to patients in clinical trials. *Stat. Med.* 1995; **14**:2661–2674.

Shao J; Tu D. *The Jacknife and the Bootstrap.* New York: Springer; 1995.

Sharp SJ; Thompson SG; Altman DG. The relation between treatment benefit and underlying risk in meta-analysis. *BMJ* 1996; **313**:735–738.

Sharp SJ; Thompson SG. Analysing the relationship between treatment effect and underlying risk in meta-analysis: Comparison and development of approaches. *Stat. Med.* 2000; **19**:3251–3274.

Shuster JJ. *Practical Handbook of Sample Size Guidelines for Clinical Trials.* Boca Raton, FL: CRC, 1993.

Simpson JM; Klar N; Donner A. Accounting for cluster randomization: A review of primary prevention trials; 1990 through 1993. *Am. J. Public Health* 1995; **85**:1378–1383.

Smeeth L; Haines A; Ebrahim S. Numbers needed to treat derived from meta-analysis—sometimes informative; usually misleading. *BMJ* 1999; **318**:1548–1551.

Smith GD; Egger M. Commentary: Incommunicable knowledge? Interpreting and applying the results of clinical trials and meta-analyses. *J. Clin. Epidemiol.* 1998; **51**:289–295.

Smith GD; Egger M; Phillips AN. Meta-analysis: Beyond the grand mean? *BMJ* 1997; **315**:1610–1614.

Smith TC; Spiegelhalter DJ; Parmar MKB. Bayesian meta-analysis of randomized trials using graphical models and BUGS. In: DA Berry and DK Stangl, eds. *Bayesian Biostatistics.* New York: Marcel Dekker, 1996, pp. 411–427.

Snee RD. Validation of regression models: Methods and examples. *Technometrics* 1977; **19**:415–428.

Sox HC; Blatt MA; Higgins MC; Marton KI. *Medical Decision Making.* Boston: Butterworth and Heinemann, 1988.

Spiegelhalter DJ. Probabilistic prediction in patient management. *Stat. Med.* 1986; **5**:421–433.

Sterne JAC; Smith GD; Cox DR. Sifting the evidence—what's wrong with significance tests? Another comment on the role of statistical methods. *BMJ* 2001; **322**:226–231.

Still AW; White AP. The approximate randomization test as an alternative to the F-test in the analysis of variance. *Brit. J. Math. Statist. Psych.* 1981; **3**:243–252.

Stockton CW; Meko DM. Drought recurrence in the Great Plains as reconstructed from long-term tree-ring records: *J. Climate Appl. Climatol.* 1983; **22**:17–29.

Stone M. Cross-validatory choice and assessment of statistical predictions. *JRSS B* 1974; **36**:111–147.

Su Z; Adkison MD; Van Alen BW. A hierarchical Bayesian model for estimating historical salmon escapement and escapement timing. *Can. J. Fish. Aquat. Sci.* 2001; **58**:1648–1662.

Subrahmanyam M. A property of simple least squares estimates. *Sankha* 1972; **34B**:355–356.

Sukhatme BV. A two sample distribution free test for comparing variances. *Biometrika* 1958; **45**:544–548.

Suter GWI. Abuse of hypothesis testing statistics in ecological risk assessment. *Hum. Ecol. Risk Assessment* 1996; **2**:331–347.

Teagarden JR. Meta-analysis: Whither narrative review? *Pharmacotherapy* 1989; **9**:274–284.

Therneau TM; Grambsch PM. *Modeling Survival Data.* New York: Springer, 2000.

Thompson SG. Why sources of heterogeneity in meta-analysis should be investigated. *BMJ* 1994; **309**:1351–1355.

Thompson SK; Seber GAF. *Adaptive Sampling.* New York: John Wiley & Sons, 1996.

Thorn MD; Pulliam CC; Symons MJ; Eckel FM. Statistical and research quality of the medical and pharmacy literature. *Am. J. Hosp. Pharm.* 1985; **42**:1077–1082.

Tiku ML; Tan WY; Balakrishnan N. *Robust Inference.* New York: Marcel Dekker, 1990.

Toutenburg H. *Statistical Analysis of Designed Experiments*, 2nd ed. New York: Springer-Verlag, 2002.

Tribe L. Trial by mathematics: precision and ritual in the legal process. *Harvard L. Rev.* 1971; **84**:1329.

Tsai C-C; Chen Z-S; Duh C-T; Horng F-W. Prediction of soil depth using a soil-landscape regression model: A case study on forest soils in southern Taiwan. *Proc. Natl. Sci. Counc. ROC(B)* 2001; **25**:34–39.

Tu D; Zhang Z. Jackknife approximations for some nonparametric confidence intervals of functional parameters based on normalizing transformations. *Comput. Stat.* 1992; 7:3–5.

Tufte ER. *The Visual Display of Quantitative Information.* Cheshire, CT: Graphics Press, 1983.

Tufte ER. *Envisioning Data. Graphics Press.* Cheshire, CT: Graphics Press, 1990.

Tukey JW. *Exploratory Data Analysis.* Reading, MA: Addison-Wesley, 1977.

Tukey JW. The philosophy of multiple comparisons. *Stat. Sci.* 1991; **6**:100–116.

Tukey JW; McLaughlin DH. Less vulnerable confidence and significance procedures for location based on a single sample; Trimming/Winsorization 1. *Sankhya* 1963; **25**:331–352.

Tversky A; Kahneman D. Belief in the law of small numbers. *Psychol. Bull.* 1971; **76**:105–110.

Tyson JE; Furzan JA; Reisch JS. and Mize SG. An evaluation of the quality of therapeutic studies in perinatal medicine. *J. Pediatr.* 1983; **102**:10–13.

Vaisrub N. Manuscript review from a statisticians perspective. *JAMA* 1985; **253**:3145–3147.

van Belle G. *Statistical Rules of Thumb.* New York: John Wiley & Sons, 2002.

Venn J. *The Logic of Chance.* London: Macmillan, 1888.

Vichers A; Cassileth B; Ernst E; et. al. How should we research unconventional therapies? Int. J. Tech. Assess. Health Care. 1997; **13**:111–121.

Victor N. The challenge of meta-analysis: discussion. *J. Clin. Epidemiol.* 1995; **48**:5–8.

Wainer H. Rounding tables. *Chance* 1998; **11**:46–50.

Watterson IG. Nondimensional measures of climate model performance. *Int. J. Climatol.* 1966; **16**:379–391.

Weerahandi S. *Exact Statistical Methods for Data Analysis.* Berlin: Springer-Verlag, 1995.

Weisberg S. *Applied Linear Regression,* 2nd ed. New York: John Wiley; 1985.

Welch BL. On the *z*-test in randomized blocks and Latin squares. *Biometrika* 1937; **29**:21–52.

Welch GE; Gabbe SG. Review of statistics usage in the *American J. Obstetrics and Gynecology. Am. J. Obstet. Gynecol.* 1996; **175**:1138–1141.

Westfall DH; Young SS. *Resampling-Based Multiple Testing: Examples and Methods for p-Value Adjustment.* New York: John Wiley & Sons, 1993.

Westgard JO; Hunt MR. Use and interpretation of common statistical tests in method comparison studies. *Clin. Chem.* 1973; **19**:49–57.

White SJ. Statistical errors in papers in the *British J. Psychiatry. Br. J. Psychiatry* 1979; **135**:336–342.

Wilkinson L. *The Grammar of Graphics.* New York: Springer-Verlag, 1999.

Wilks DS. *Statistical Methods in the Atmospheric Sciences.* New York: Academic Press, 1995.

Willick JA. Measurement of galaxy distances. In: A. Dekel and J. Ostriker, eds. *Formation of Structure in the Universe.* New York: Cambridge University Press, 1999.

Wilson JW; Jones CP; Lundstrum LL. Stochastic properties of time-averaged financial data: Explanation and empirical demonstration using monthly stock prices. *Financ. Rev.* 2001; **36**:3.

Wu CFJ. Jackknife, bootstrap, and other resampling methods in regression analysis (with discussion). *Ann. Stat.* 1986; **14**:1261–1350.

Wulf HR; Andersen B; Brandenhof P. and Guttler F. What do doctors know about statistics? *Stat. Med.* 1987; **6**:3–10.

Yandell BS. *Practical Data Analysis for Designed Experiments.* London: Chapman and Hall, 1997.

Yoccuz NG. Use, overuse, and misuse of significance tests in evolutionary biology and ecology. *Bull. Ecol. Soc. Am.* 1991; **72**:106–111.

Yoo S-H. A robust estimation of hedonic price models: Least absolute deviations estimation. *Appl. Econ. Lett.* 2001; **8**:55–58.

Young A. Conditional data-based simulations: Some examples from geometric statistics. *Int. Stat. Rev.* 1986; **54**:1–13.

Zhou X-H; Gao S. Confidence intervals for the log-normal mean. *Stat. Med.* 1997; **17**:2251–2264.

Zumbo BD; Hubley AM. A note on misconceptions concerning prospective and retrospective power. *Statistician.* 1998; **47**:385–388.

Author Index

Adams DC, 90
Adkins NC, 105
Adkison MD, 90
Albers W, 54
Altman DG, 13, 50, 74, 75, 101, 105
Aly E-E AA, 62
Andersen B, 37
Anderson DR, 14
Anderson S, 54, 106
Anderson SL, 60, 75
Anderson W, 55
Anscombe F, 14
Archfield SA, 150
Avram MJ, 74

Bacchetti P, 75
Badrick TC, 74, 104
Bailar JC, 105
Bailey KR, 88
Bailor AJ, 60
Baker RD, 61, 70
Balakrishnan N, 60
Barbui C, 36
Barnard GA, x
Barnston AG, 162
Barrodale I, 137
Bayarri MJ, 86
Bayes T, 79
Begg CB, 105
Bent GC, 150
Berger JO, 23, 37, 41, 75, 86
Berger VW, 19, 36, 37, 50, 74, 90, 92
Berk KN, 158

Berkeley G, 75
Berkey C, 88
Berlin JA, 90 9
Berry DA, 23, 84
Berry KJ, 90, 138, 143
Bickel P, 50, 54
Bithell J, 47
Bland JM, 50, 74
Bly RW, 27, 38
Blyth CR, 23
Bock M, 164
Boomsma A, 162
Bothun G, 32, 139
Box GEP, 60, 71, 75
Brockwell PJ, 157
Browne MW, 157
Burnham KP, 14

Cade B, 138
Callaham ML, 103
Campbell MJ, 75
Camstra A, 162
Canty AJ, 78, 90
Capaldi D, 93
Carlin BP, 90
Carpenter J, 47
Carroll RJ, 50
Casella G, 37, 41, 50
Chalmers TC, 36, 88
Chernick M, 78
Cherry S, 74
Chiles JR, 13
Christophi C, 37, 92
Chu KC, 21

Subject Index

a prior distribution, 81–83
a prior probability, 86
Acceptance region, 101–102
Accuracy vs. precision, 187
Agronomy, 33, 66
Algorithms, 137, 160, 161
Allocation sequence, 36
Aly's statistic, 62–63
Analysis of variance, 62, 65
Angiograms, 93
Animals, 7
Antibodies, 26
ARIMA, 157
Arithmetic average, *see* mean
Aspirin, 35, 66
Association, 136
  spurious, 143
  versus causation, 136, 141
Assumptions, 4, 32, 51, 61, 63, 71–78,
    89, 101, 129, 148
Astronomy, 8, 74, 103
Asymptotic, 51, 165
  approximation, 60
  relative efficiency, 43–44
Audit, 139
Autocorrelation, 104
Autoregressive process, *see* time series
Axis label, 111, 122
Axis limit, 10
Axis range, 110

Bacteria, 96
Balanced design, 65
Bar chart, 108, 117

Baseline, 58, 110
Bayes
  analysis, 89
  factor, 85
  Theorem, 80–84
Behrens-Fisher problem, 54
Bias
  estimation, 103, 174
  sample, 7, 35–36, 47, 87, 102–104,
    140, 150+B397
Bias-corrected and accelerated, 47,
  61
Blinding, 36, 92
Blocks, 33, 59, 71
Blood, 9, 27, 74
Blood pressure, 26
Bonferroni correction, 72
Bootstrap, 49, 70, 90, 152, 173
  limitations, 78
  nonparametric, 45
  parametric, 48, 99
  percentile, primative, 48, 61
  sample, 29–30, 46, 57, 147+B221
  test, 51–56
  tilted, 79
Box and whiskers plot, 96–99
Boyle's Law, 5, 130

Cancer, 17, 21, 84, 87, 135, 140
Caption, 122
CART, 152, 155
Categories, 26, 71
Cause and effect, 133, 134, 155, 158
Censored data, 94, 103

Monotone function, 63, 69, 130
MRPP, 138
Multivariate analysis, 142

Negative findings, 106
Neural network, 152
Neurology, 103
Newton's Law, 20, 74
Neyman-Pearson theory, 15
Nominative label, 112
Nonresponders, 30–31, 72
Normal alternatives, 54
Normal distribution, 67, 71, 145
Normal scores, 53

Objectives, 4, 11
O'Brien's test, 55
Observational studies, 87
Observations
   dependent, 56, 148
   exchangeable, 51
   identically-distributed, 33
   independent, 45
   independent, identically distributed,
      32, 51, 101
   multivariate, 56
   transformed, 64
Odds, 85
One-sided vs. two-sided, 16–17
Ordinal data, 114, 187
Outliers, 78, 138
Overfitting, 160

Parameters, 45, 48, 51, 55, 73, 88,
      101, 131
   location, 29, 43
   scale, 29
   shift, 44
Paternity, 81
Patterns, 7, 22
Percentiles, 45, 47, 79, 99
Permutation test, 14, 19, 21, 51–56,
      59–63, 67, 70–71, 77, 89,
      145+B397
Permutations, 69, 138
Perspective plot, 115, 117
Phase III trials, 23
Physician, 34
Pie chart, 117, 118, 125
Pitman correlation, 64

Pivotal quantitty, 49, 57, 78
Placebo, 35
Plotting region, 110, 122
Plotting symbol, 118, 122, 123
Poisson regression, 148
Poisson variable, 147
Poker, 9
Polar coordinates, 117
Polyline, 114, 118, 119
Polynomial, 132
Population, 3–5, 25, 31–33, 74
Population statistics, 5
Poverty, 134
Power, 16, 28–30, 54, 55, 63, 72,
      85
   post-hoc, 104
   related to sample size, 28, 91
Precision, 46, 95, 97, 137, 139
Prediction, 140, 159–161, 173
Pricing, 103
Proc ARIMA, 138
Proc Means, 53
Proc Mixed, 58
Protocol, 8, 13, 87
Proxy variable, 135
p-value, 10, 54, 71–75, 100–105

Radioimmune assay, 130
Rain, 156
Random number, 7, 98
Randomizing, 33–36, 54, 84
Ranks, 52
Ratio, 47
Raw data, 90
Recruitment, 27, 36
Recruits, 6, 13, 34
Redheads, 12, 15, 34
Redshift, 104
Regression
   coefficients, 137–142
   linear vs. nonlinear, 131
   multivariable, 145, 147
   scope, 129
   stepwise, 145
Regulatory agency, 15, 35
Rejection region, 16
Relationship, 130
Relativity, 20, 74
Repeated measures, 56
Report, 26